THE ETHICS OF THE GLOBAL ENVIRONMENT

UNIVERS
LEARNING
ONE WEEK L

EDINBURGH STUDIES IN WORLD ETHICS

Other titles in the series:

World Ethics: The New Agenda
Nigel Dower

Ethics, Economics andd International Relations
Peter Brown

Development Ethics
Des Gasper

THE ETHICS OF THE GLOBAL ENVIRONMENT

Robin Attfield

EDINBURGH UNIVERSITY PRESS

© Robin Attfield 1999

Edinburgh University Press
22 George Square, Edinburgh

Typeset in Times
by Hewer Text Ltd, Edinburgh and
printed and bound in Great Britain by
MPG Books, Bodmin

A CIP record for this book is available
from the British Library

ISBN 0 7486 0895 8 (paperback)

The right of Robin Attfield
to be identified as author of this work
has been asserted in accordance with
the Copyright, Designs and Patents Act 1988.

11800380

UNIVERSITY OF GLAMORGAN
PRIFYSGOL MORGANNWG
Learning Resources
Centre

CONTENTS

ACKNOWLEDGEMENTS

I should like to thank the following for their helpful suggestions concerning the text of this work: David Attfield, Nigel Dower, Markku Oksanen, Neil Thomas, and an anonymous referee of Edinburgh University Press, and also the Oxford Centre for Environment, Ethics and Society, to a Seminar of which most of Chapter 3 was presented. Seminars of the Global Citizenship Project of the University of Aberdeen Centre for Philosophy, Technology and Society proved influential in the composition of Chapter 11. But responsibility for the final text and its shortcomings remains my own.

I am most grateful to Robin Wackerbarth for his assiduous checking of successive drafts, and for his earlier contributions to a customised database, ingeniously devised by Leela Dutt, and also to Mary Jo Slazak of Rowman & Littlefield and to the Library staff of University of Wales Cardiff, and in particular to Tom Dawkes, for invaluable bibliographical assistance. Thanks are further due to University of Wales Cardiff for releasing me from other duties for one Semester to complete this work, and also to my colleagues of the Philosophy Section for undertaking those duties. Above all, thanks are due to my wife, Leela Dutt, without whom this work would not have been begun, continued or completed.

LIST OF ABBREVIATIONS

BATNEEC	Best Available Technology Not Entailing Excessive Cost
CA	California
CFCs	chlorofluorocarbons
CITES	Convention on International Trade in Endangered Species
FAO	Food and Agriculture Organisation
GA	Georgia
GATT	General Agreement on Tariffs and Trade
GDP	Gross Domestic Product
GNP	Gross National Product
HCFCs	hydrochlorofluorocarbons
HMSO	Her Majesty's Stationery Office
ICRW	International Convention for the Regulation of Whaling
IMF	International Monetary Fund
IPCC	Intergovernmental Panel on Climate Change
IUCN	International Union for the Conservation of Nature
MA	Massachusetts
MD	Maryland
MI	Michigan
MIT	Massachusetts Institute of Technology
NEEC	Not Entailing Excessive Cost
NGOs	Non-governmental Organisations
NJ	New Jersey
UK	United Kingdom
UNCED	United Nations Conference on Environment and Development
UNEP	United Nations Environment Programme
USA	United States of America
USSR	Union of Soviet Socialist Republics (former)

VT	Vermont
WCC	World Council of Churches
WCED	World Commission on Environment and Development
WCU	World Conservation Union
WDM	World Development Movement
WHO	World Health Organisation
WWF	World Wide Fund for Nature

INTRODUCTION

As seen from space, our planet, the shared natural environment of humanity and fellow creatures, is both valuable and vulnerable. It is valuable as the bearer and the setting of valuable lives, and vulnerable through the maltreatment by its inhabitants of shared resources, and of each other. In circumstances such as these, a global ethic relevant to the environment, and applying both to individuals, institutions and countries, becomes indispensable. So too is its study, global ethics, the subject of this book.

There are some who disparage ethics and values, urging identification with nature instead; once we become identified with the natural environment, they say, the necessary action will become obvious, if not instinctual. But such attempts to merge the individual with nature forget that it is largely as individuals with distinct identities, or as groups of such individuals, that we think and act, and that we need to respect as other than ourselves the people and creatures around us, and the rest of the natural world, as we interact with them. Only on this basis is a sense of solidarity, or even of belonging, a possibility. They also forget that, for actions to have reasons, values are needed, and that if values are to be sifted and prioritised, ethics really is indispensable.

The first part of the book seeks to clear the ground for the study of the ethics of global environmental problems. The first chapter attempts to analyse and clarify the concepts of nature and environment, which are much less clear than they may seem. Human beings, for example, interact with nature, and yet are also part of nature. There again, they sometimes resolve to regard as natural only what is uninfluenced by humanity, and then discover that on the surface of our planet almost everything betrays a human touch. And if the environment seems less problematic, the question soon arises of whether this does not just comprise the surroundings (or perhaps the significant surroundings) of each human being, and of whether, if so, talk of the global environment (the kind of talk present in the opening of this Introduction) even makes sense. This chapter defends the concepts of the global environment, and argues for shared national and international responsibilities in its regard. While readers anxious to plunge into ethics could move ahead to the next chapter, those willing to reflect on key concepts relating to the global environment, and pivotal for many of the issues discussed in later chapters, should begin here.

Chapter 2, in keeping with other works in the World Ethics series, contrasts three approaches to ethics, realism, communitarianism and cosmopolitanism, surveys key stages in the history of the third of these, and defends both this kind of approach and a particular form of it, biocentric consequentialism. Cosmopolitanism refuses to draw ethical boundaries where obligations are concerned, and is here argued to cope with problems of ethics (including global ethics) in ways unmatched by the other approaches. Consequentialist theories are preferred because all the foreseeable consequences of action (and inaction) are taken into account, and biocentric theories because they include all bearers of interests, and not human interests only. Some of the problems for these theories are considered in later chapters.

In the third chapter, the tradition which regards humanity as trustees of the natural environment of the planet is discussed and upheld against a wide range of criticisms. The shared belief of Judaism, Christianity and Islam that human beings are stewards answerable to God for the care of nature is defended; also a secular version of the stewardship (or trusteeship) approach, including a secularised understanding of answerability, is argued to be coherent, and open to people without any form of theistic belief. The trusteeship approach is, in my view, consistent with a range of ethical positions, including that defended in the previous chapter (biocentric consequentialism). By providing a credible metaphysical backdrop for the role of humanity (whether religious or secular), it also reinforces motivations for following such an ethic, supplementing the reasons for action which this ethic enshrines such as concern for the good of human beings and fellow creatures.

The fourth chapter comprises thought-experiments concerning the possibility of human extinction. This subject is studied not as if extinction were a serious likelihood, but because reflection on its possibility serves to elicit values which are often not consciously recognised. It emerges that we have like reason to care about future lives as about present ones, and also that we may well have reasons of self-interest to care about something larger than ourselves. The findings of this chapter prove to cohere with the ethic defended earlier, which supplies reasons for caring, including caring about future generations, as well as supplying guidance about policies and conduct in the present. They also cohere with the trusteeship approach.

Part Two applies the ethic of Part One to global environmental issues, beginning in Chapter 5 with issues surrounding resources, affected as they now are by the problem of global warming. Contemporary problems relating to forests, energy and water are introduced and discussed, interim conclusions are drawn, and principles are presented for the evolution of the post-Kyoto international regime for the regulation of greenhouse gas emissions.

Chapter 6 discusses and defends the concept of sustainable development. While this concept is sometimes employed as a mere rhetorical device, and is sometimes used to justify policies imposed by Northern (or 'developed') countries on Southern ('developing' or Third World) countries, development of a sustainable kind is argued to be vital for the satisfaction of current needs, and for the sake of the foreseeable future, as was recognised at the Rio Conference on Environment and Development in 1992. The aim is not to assimilate Third World consumption to current American levels, but to attain sustainability worldwide on an equitable basis. Without sustainable development, environmental problems will predictably worsen. However, its attainment involves a restructuring of international trade and finance, and thus of international relations.

While environmental problems are not principally due to population growth (as opposed to global technological change on the one hand, and poverty on the other), the problems of population and of poverty are discussed in Chapter 7 against the background of the need to feed a human population on a sustainable basis. If humanity is to have adequate nutrition, a stabilisation level of around the eight billion target envisaged at the UN Cairo Conference of 1994 (rather than a much higher level) becomes crucial, as do related national population policies for all countries. Despite its questionable use of coercion, the Chinese population policy may, I claim, comprise an important contribution.

Sustainability, however, involves the preservation of the ecosystems on which humanity and other creatures depend, and thus of biological diversity, including the preservation of most of the species of the planet. Chapter 8 discusses issues surrounding biodiversity preservation, in the light of the Biodiversity Convention of the 1992

Rio United Nations Summit on Environment and Development. Consequentialism is found helpful in determining how strictly policies of preservationism should be followed. Principles for equitable international funding for biodiversity preservation are also discussed, as are principles for the recognition of indigenous knowledge among Third World peoples.

Part Three steps back from the problems to relate the emerging conclusions to an understanding of global justice, of international order, of equity across the generations, and of global citizenship. Chapter 9 compares different accounts of global justice, and bears out the importance of a theory which is biocentric as well as consequentialist. It also discusses regimes for the global commons (such as the oceans, the atmosphere and Antarctica), and the desirability of an international order short of world government in which sovereignty is to some degree pooled in international institutions.

Chapter 10 discusses the importance of recognising diverse national perspectives on global policies aiming at sustainability, in view of the need for further international negotiations to develop the regime of Kyoto (1997), in the light of the largely successful agreement (Montreal, 1987, amended at London in 1990) to ban CFCs. It also sifts principles of intergenerational equity, some of which prove vital for sustainability, and gives special consideration to the Precautionary Principle (which urges intervention to prevent possible disasters in advance of the availability of scientific information), its justification and its scope.

The final chapter relates consequentialist obligations to campaigning for political change (a possibility which in my view helps to overcome a recurrent objection to consequentialism), and investigates whether enhanced procedures for decision-making are desirable or necessary for environmentally sensitive decisions. It argues for the appointment to legislatures of a small number of proxies to represent future generations and non-human creatures, but claims that environmentally sensitive decisions and policies are possible in apparently unpromising contexts, and need not await procedural or constitutional reforms. It also defends a concept of global citizenship which does not presuppose a global state, but involves participation in one or another of the worldwide networks comprising global civil society. Global civil society, itself an aspect of globalisation, is capable of challenging other aspects, and of enhancing prospects for global solutions.

The ethical theory presented in this book is more fully defended in *Value, Obligation and Meta-Ethics* (Amsterdam and Atlanta, GA:

Rodopi, 1995), and applied to environmental ethics in *Environmental Philosophy: Principles and Prospects* (Aldershot: Avebury and Brookfield, VT: Ashgate, 1994). In the present book it is newly applied to global and international issues. Every effort has been made to keep abreast of conventions, institutions and conferences (for which the List of Abbreviations may be found helpful by some readers); but unexpected developments arising after mid-1998 could require some passages to be imaginatively updated by observant students of international affairs. But the ethical principles defended here, and concepts such as those of the global environment, global citizenship and global civil society are, happily, unlikely to age, and can safely be commended to readers of the third millennium as well as those of the final year of the second.

PART I
CONCEPTS, THEORIES AND VALUES

CHAPTER 1

NATURE AND THE GLOBAL ENVIRONMENT

What kind of ethic and what kinds of international action are needed to tackle environmental problems such as global warming, the growing gaps in the ozone layer, the destruction of rainforests, the growth of deserts and the pollution of the oceans? These problems concerning the global environment are among the issues to be addressed in later chapters of this book, issues also including resources and resource consumption, sustainable development, population and poverty, the preservation of biodiversity and global justice. But before they can be addressed, the concept of the global environment needs to be clarified, in view of criticisms that it is a confused concept, and that people concerned about environmental problems need to start somewhere else. Elucidating this concept involves, in part, asking just what an environment is, and also what gives a global one its global character.

ENVIRONMENTS AND THE GLOBAL ENVIRONMENT

There are countless environments, and yet belief in *one* global environment seems inescapable. 'Essentially an environment only exists because it is inhabited by a [particular living] organism. Thus a field is the environment for a cow, a cow-dung pat is the environment for a dung-beetle, and the exoskeleton of the dung-beetle is the environment of a parasitic mite. Therefore the field comprises an infinity of overlapping environments'[1] Given this sense of 'environment', deriving from the biological sciences, each human group and individual has its own environment, and each forms part of the environment of many of the others. 'Environment' here means (roughly) 'encompassing system'.

Yet, according to the same author, ' "Environment" is also used in the sense of . . . ecosystem' (a system of interacting living organisms and non-living elements), and 'since an ecosystem is usually thought of as occurring within a self-contained and restricted area, and since complete isolation of most areas is impossible, it can be argued that the Earth itself is the only real ecosystem'.[2] This is no less true when the environment is not a single ecosystem, but the network or system of such systems. Much the same holds good when 'the environment' is used not of ecosystems at all, but of the natural world as a whole, except that the natural world includes the stars and planets as well as the Earth.

So there are tensions generated by the very concept of environment; and these tensions will be relevant when we tackle environmental problems and their solutions. For many writers, environments are necessarily relative to the creatures environed. David Cooper, for example, reminds his readers of the concept of environment as *milieu* or field of significance, an area which an individual animal (such as a badger) or an individual human being knows its way around and cares about. He contrasts this existential concept with the detached scientific concept of an environment as a causal system, and urges environmental philosophers to focus on the existential concept, rather than on what he seems to regard as the inflated causal concept of the global environment. Intellectuals, he holds, are prone to be at home everywhere, but at the same time nowhere in particular; hence their preference for such a grandiose notion. Few loyalties are likely to be attached to the planetary biosphere, and those who advocate environmentalism of the global kind cannot help underselling even those ethical issues, such as animal welfare and justice for the Third World, which they may associate with this very cause. But those who care about their own (often local) arena of significance may eventually come to care about others' environments (in the same sense), and make common cause with theirs.[3] This apparently plausible view warrants a reply; but first another relational view should be considered.

For a subtly different perspective has been offered by Tim Ingold, a perspective which seeks to disclose the pre-ethical commitments which we all have to our environments, in virtue of 'the necessary situatedness of human beings within the context of their active, practical and perceptual engagement with their surroundings'.[4] Environments never exist before the environed creature does, and cannot exist without such a creature. They comprise a process rather than a fixed objective entity, and are continually under construction through the activities of the living being environed. Hence a distinction should be made between environment and nature, and we should be wary of expressions such as 'the natural environment'. For while nature is a world which exists apart from ourselves, and can be studied in a detached, scientific manner, and is given in advance of human history, environments are fundamentally historical, cannot be understood with scientific detachment, and belong to the same meaningful lifeworld as ourselves. All kinds of distortions arise for ethics when the environment is instead conceived as the preconstituted base of human action, rather than a world within which people dwell.[5]

The distortions in the field of ethics alleged by Ingold can be deferred until the chapter on environmental ethics is reached; but this is the best place to consider his account of pre-ethical commitments. We should at once notice, however, that his account of our pre-ethical engagement with our surroundings ends with this being described as a condition 'by virtue of which we are all fellow passengers on this planet of ours'.[6] Ingold seems to diverge here from Cooper, for his words effectively grant that the entire planet can be or become our shared environment (in the meaningful sense). While neither Cooper nor Ingold requires environments to be local, both seem to confine them to the segments of the world which an individual actually experiences; but Ingold seems to recognise that this would too severely restrict what people can care about, and that each person can come to care about the global environment which she and other creatures share. Even Cooper could imaginably be understood as recognising this too, in holding that individuals involved in 'pockets of resistance' can empathise with others concerned for other such pockets, despite his disparagement (in the same breath) of 'global awareness'.[7] In any case (whether or not these are wishful interpretations of Ingold and Cooper), the extension of a field of significance from local to global is not impossible, since it happens quite often, as recent increases of the membership of campaigning NGOs such as Friends of the Earth and Greenpeace eloquently attest.

However, the neighbourhood sense of 'environment' should not be allowed to predominate over other senses. While there is such a sense, and an environment can be regarded as an experiential and interactive process rather than a causal system, there are other senses of 'environment' which make individuals vulnerable to environmental factors such as the hail now beating on my window, and make us dependent on our environments, not least for air, food and drink, rather than continually constructing them and their meaning. Otherwise, trees and other non-conscious creatures would not have environments at all, lacking perceptions and intentions as they do; and ecologists could not define 'environment' in any of the ways mentioned at the beginning of this chapter.

THE CONCEPT OF ENVIRONMENT

As Nigel Dower has written (in response to Cooper), we have both intentional (perspective-dependent) and also objective concepts of environment. Objective concepts include the concept of the environment as an objective system of causes and effects. This concept too can be a relational concept, in that entities which exert a causal influence on something are clearly related to it; but the relation is different, and can be quite independent of awareness and understanding on the part of an environed subject. Nonetheless, as an object of understanding, a causal or objective environment (such as a rainforest) or its components (trees, orchids, snakes and insects, say) may be recognised as bearers of value, whether intrinsically or otherwise, and it (and they) may thus come to be cherished just as much as a familiar neighbourhood can be, though for different reasons. Such an environment will usually be the shared environment of many people and other creatures, underpinning the fields of significance of them all; so it comprises an interpersonal environment.

A conscious subject may thus have an environment in at least two senses, and it is important that, as individuals, we can employ both the intentional (field of significance) sense and the interpersonal (objective system) sense, sometimes using them about the same complexes of entities. It is also important that we can shift between these senses, not least because the concept of environment and similar concepts (such as 'sphere' and 'world') are tailored to allowing us to do so;[8] intentional and interpersonal spheres impinge on one another too much for things to be otherwise. So if pollution threatens our favourite places, we can seek, by appeal to environmental preservation, to protect our own familiar haunts, the local landscape and the national (or world) heritage at one and the same time; and we can find allies among all those opposed to pollution, as well as among others who frequent the same nooks and crannies.

It is in any case no coincidence that people possessed of the intentional or existential concept of environment, as neighbourhood, also make tacit appeal to the objective concept of environment, as encompassing system. As Andrew Belsey has pointed out, concern for a neighbourhood, or for our children's and grandchildren's inheritance, is condemned to futility, in face of global problems such as the greenhouse effect, the loss of the ozone layer, the pollution of the oceans and the disposal of nuclear waste; such concern is manifestly futile if it is restricted to our neighbours and their attitudes and behaviour. For these (and other comparable problems such as the destruction of rainforests and the growth of deserts) are global problems, involving damage to global systems, and requiring nothing less than global solutions.[9] Their very description involves reference to a global environment, shared by neighbours and non-neighbours alike. Further, as Dower remarks, there could not be neighbourhoods or fields of significance at all if there were no common environment supporting them all, and all the people who participate in them; thus 'concern for the quality of the perceived environment . . . cannot help but be mediated by what we do for this one common publicly shared environment which causally underpins all those shared environments'.[10] To put matters another way, we all, rich and poor alike, have little or no choice but to take notice of and to care about the global environment when the actions of people thousands of miles away begin to change our local countryside and climate, or (come to that) we theirs.

So, as it turns out, environments need not, after all, be fundamentally historical, or incapable of detached, objective study, or even have a significance to which we are committed before we reflect on them. Nor, where the environment is a causal system, is our situatedness in our particular environment necessary and inevitable; we are of necessity situated somewhere, but particular environmental factors and impacts are contingencies (and many of them, fortunately, alterable). Environments importantly can and do exist before the creatures they eventually environ, and as such are not invariably processes even partly constructed by the activity of those creatures. Nor are they invariably fields of significance, which we know our way around (or the world would not be such a puzzling place as it is). Besides, the distinction suggested by Ingold between the environment and nature now becomes questionable, for the distinction does not arise where an environment is an interpersonal tract of nature. As Dower comments, we can change our environment, either through physical intervention (particularly where the environment in question is such an interpersonal tract), or alternatively (in the case of perceived environments) through ourselves changing or undergoing a change of perspectives.[11] Only in the second sort of case does Ingold's distinction remain significant.

It remains important that most people do have fields of significance and spheres of perceived meaningfulness, and are thus liable to care about one environment or another; and this is one of several factors which suggest that reflective environmental ethics is not invariably destined to fail to strike a responsive chord, nor always to prove futile because people lack the loyalties, commitments and other motivations needed to translate responses into effective action. Yet Cooper's view that we cannot help but feel part of our environment, provided that we still have one at all, is unduly optimistic.[12] For environments, like homes, vary in quality (whichever sense of 'environment' we elect to use). While some are welcoming, others are alienating, or at least unsuited to love and loyalty, and many require repairs or reconstruction. By the same token, participation in preservative or restorative effort is not always self-motivating, but people can often become motivated to the necessary effort by a sense of participation in a common cause, shared in sometimes by people whose roots lie in distant settings or even in alien ground.

Dower makes a related point. Fields of significance are prone to be too narrow and confined when contrasted with the problems affecting the environment as a causal system. There is often a 'mismatch' between the level of cherished neighbourhoods and the levels at which change (including political change) is going to be needed. Sometimes the quest for 'quality of life' within a local or familiar terrain even contributes to the deterioration of the wider environment (as in the Not-In-My-Back-Yard syndrome); and equally often, the continued existence of fields of significance is itself threatened by the worsening state of the encompassing causal system. More positively, our ability to achieve changes of practice which prevent such widespread deterioration depends on expanding the extent of our fields of significance both in space and in time. The saying of John Donne in the seventeenth century shows a profound grasp of the same theme: 'No man is an island, entire of itself; . . . I am involved in mankind. And therefore never seek to know for whom the bell tolls; it tolls for thee.'[13]

The contemporary challenge is to make the planetary environment, or such broad tracts of it as the forests or the seas or generally the next generation's inheritance, become fields of significance, both for ourselves and for others. Dower here adjusts the words of the poet Piet Hein, so as to read: 'We are global citizens with local souls', and suggests that 'we need to acquire global souls' to match our global citizenship.[14] Without claiming already to have a fully formed sense of global citizenship (a topic to be addressed in the final chapter of this book), or assuming its presence in others, we can recognise Dower's remark as an expressive way of imparting that environmental ethics can suggest change on the part of its participants and their sense of identity, as well as in international relations and the structure of the world system. As Belsey points out, the context of many environmental problems is a global one, and our concepts must be no less extensive in their scope.[15]

Dower stresses that a wide range of perceived environments or fields of significance is to be foreseen and expected, and their diversity welcomed; what is important is that the various fields of significance be sufficiently enlarged to be capable of bringing about the preservation or restoration of the shared, objective environmental base. No uniform perception of or commitment to the planet is in question. He adds that not just any attitudes which are global in scope will contribute to global solutions; for exploitative global attitudes are no better than having none at all.[16] Indeed, attitudes need to be informed by reflection on principles of the right general kind, and to take account of the very diverse situations (and environments) of different people and their different scope and capacity for action. In the rest of this book I hope to throw light on how best to think about these global issues and how to confront them. The suggestion is not that changed individual attitudes would be sufficient (as opposed to necessary) to solve the problems; for political, economic and social structures also need to change. But that is all the more reason both for reflection and for the kinds of change advocated being comprehensive.

NATURE AND THE NATURAL ENVIRONMENT

As mentioned at the outset, the environment is often thought of simply as the natural world; and we have found reasons to doubt the distinction suggested by Ingold between these two concepts, except when what is in question is someone's perceived environment. Like the environment, the natural world precedes and transcends our existence, comprises an encompassing system, open to scientific study, and sustains the significant places of all our lives. Unlike the global environment, the natural world includes the solar system and the galaxies. But at the terrestrial level, nature simply is the natural environment, as opposed to the social environment, the sphere of human culture, and the built environment, its most striking expression.

Attempts to define nature as tracts of the universe unaffected by humanity are by this stage in history unhelpful, leading to the conclusion of Bill McKibben (who employs a definition of this kind) that nature is now extinct, at least on Earth, as the imprint of humanity is spread across the entire planet.[17] Yet natural laws continue to operate; the natural evolution of species has not been curtailed, biotechnology notwithstanding; and our own bodies continue to be dominated by the natural cycles of days, months and years. In other words, McKibben's definition of 'nature' does not fit many of our uses of that concept. Besides, theorists who deny value to tracts of nature modified by humanity, however slightly (as no longer 'natural'), have to account for people's widespread love of places such as mountain trails, streamside walks, beaches and the sea; for these are just the kinds of places widely cherished as natural settings or as the natural environment. (Naturalness may not in any case be the fundamental ground of value; but that is a separate issue.)

It would certainly be misleading to equate nature with the entire material universe in all its aspects, as this would elide the distinction between nature and culture. But the distinction between nature and culture can be preserved if nature is understood as tracts and processes not predominantly modified or shaped by human activity. In this sense both nature and the natural environment undeniably remain in being. What is more, all human beings depend on them, and all other living creatures too.

The definition of 'nature' just given in no way precludes recognition that the natural world (and thus the natural environment) has been significantly affected by humanity, and thus that nature itself, by now, has a history. For example, some more or less stable ecosystems (such as the Long Mynd in the Welsh Marches of England) have depended for centuries on being grazed by sheep, and ecologists now seek to preserve the plant and insect species of this modified environment, as opposed to those which preceded human occupation. Indeed, most of the countryside of Europe is either tilled, grazed or afforested, and the face of the landscape has considerably changed several times over through historical processes like the enclosure movement and through the replacement of horses with tractors in the twentieth century. Yet the preservation of this same countryside remains the aim of voluntary nature-conservation bodies such as the Council for the Preservation of Rural England and official bodies such as English Nature. While countrysides are sometimes so manicured as to belong to the sphere of culture rather than that of nature, it would be misguided to classify countryside as no part of nature for such reasons as that it is not wilderness and is partly a product of culture.

But if farmed landscapes can be regarded as areas of nature, the suggestion might seem to be worthy of consideration that nature itself has effectively become a social construct. Appeals to *human* nature, someone might argue, often tacitly amount to condonements or prescriptions of particular kinds of conduct, expressing thereby a socially constructed view of human nature; and it is sometimes suggested that nature itself is socially constructed too, forming a projection of society's attitudes concerning what it expects to find the world to be like.[18] Arguably, claims about institutions such as the state or heterosexual marriage being 'natural' make better sense if understood in this way, or at least as having implications about what ought to be (normative implications, that is). But the normative sense of 'natural' (meaning something between 'normal', 'acceptable' and 'best') needs to be distinguished from the sense of 'natural' as used here, that of 'pertaining to tracts and processes of the material world not predominantly modified or shaped by human activity'. This second sense will continue to have an application as long as study of the world of nature continues; and there is no need to grant that nature in this sense is constructed by society.[19] Besides, as argued above, there has to be an objective, interpersonal natural environment if there are to be subjects or selves with perceptions and with intentional environments, or indeed if human society is to exist at all. The suggestion that nature is and always was a construct of society is an impossible claim, which can only be made if, contrary to what is claimed, there is a natural system of causes and effects (a system, that is, not originated by humanity) on which society itself depends, and on which we ourselves depend. Importantly, then, nature is an objective causal system, and not a social construct.

This understanding of nature, however, leaves considerable scope for different conceptions of nature, or ways in which the objective system of nature is understood. Rival extreme views, for example, regard nature as respectively a living organism or as a mechanism, while other views stress how only some tracts of nature can be regarded in either of these ways, much of nature being neither alive nor mechanical. (The Gaia hypothesis regards the Earth as a self-restoring organism,[20] but has not found favour among more than a minority of qualified scientists.) Further views make nature either a bottomless mine of precious resources, or a bottomless sink for the absorption of pollution; as will be seen in later chapters, these are misconceptions, which sometimes exacerbate environmental problems. They also clash with the view of nature as a sanctuary or temple, a fitting view, perhaps, for particular settings such as the Grand Canyon. But it is impossible so to regard nature as a whole, unless we abandon reliance on nature's assimilative capacity and on natural resources. Yet we are simply unfree to abandon this reliance, whatever our wishes in the matter may be.

Another conception of nature concerns its wildness, and generally its otherness. While nature supplies our home, it is never so tame as simply to comprise a home. As Henry David Thoreau writes, 'We need the tonic of wildness.'[21] Not being planned or devised by humanity, wild nature can remind us of the pettiness of our parochial concerns, or restore our sense of proportion, when we perceive the very indifference of its processes to our own machinations, or when we catch sight of the alien nature of a wild, independent natural creature such as a kestrel, as depicted by Iris Murdoch,[22] or such as a falcon, lauded in 'The Windhover' by Gerard Manley Hopkins,[23] and in *Walden* by Thoreau.[24] Indeed this recognition of nature's otherness is not easily reconciled to the view of environments as simply familiar haunts and pathways; besides, awareness of nature's otherness is premised on the natural environment being there to be discovered, transcending our consciousness, and by no means our creation. Yet nature has also been found to embody regularities, and is to that extent comprehensible; and it is so far from being uniformly wild that cultivation and domestication are possible, without natural processes and interactions being abolished thereby. Experience of nature's otherness is probably needed for human renewal and refreshment; yet these needs can be provided for to a remarkable extent through gardens, parks, countryside trails and nature reserves, and the provision of these experiences of nature for countless citizens of the modern world would be impossible if nature were wholly other or wild.

Diverse conceptions of nature thus correct one another. One potentially dangerous conception is of nature as a stock of natural capital, of which the economic value is to be kept constant. Besides extending the scope of human business enterprises to the bounds of the universe, this approach is inconsistent with anything in nature having value independent of human attitudes and interests. Certainly, the contrary view that the whole universe has such value seems far-fetched when applied to abiotic nature, for example in the form of interstellar dust; but the belief that nothing outside the realm of culture has such value is equally difficult to defend (see Chapter 2). In any case, the evolutionary process has to be regarded as the matrix of all value (unless, that is, non-living matter can have independent value), since it has facilitated both human life and culture, and the ways of life of all other living species and of the ecosystems in which they participate. While this does not mean that this process has an independent value of its own, the view that each and every non-human living creature is in principle expendable in the cause of human (and particularly of commercial) interests is manifestly absurd.

Similar considerations are relevant to the question of whether nature can be owned, and if so, by whom. While portions of nature are plausibly ownable, it has been argued, as by Gunnar Skirbekk and his fellow authors of *The Commercial Ark*, that the biosphere (the sphere which encircles our planet and consists of living creatures and of the elements on which they depend) cannot be owned at all,[25] while Karl Marx maintained that not even all the societies of the current generation can own the land.[26] The absurdity of any attempt to imagine how the biosphere might be owned conveys that there are limits to the ownership of nature; this is not the place to resolve whether these are simply physical limits, or moral limits too. These issues will be further considered in the context of the stewardship tradition in Chapter 3.

We have seen that, except where perceived or intentional personal environments are in question, the natural environment is simply an encompassing portion of nature on our planet, where nature consists in tracts and processes of the material world which have not been predominantly modified or shaped by human activity (for example, in the ways that cities and freeways have been). Nature (in this sense) is not a social construct, despite widespread tendencies to attach normative implications to the use of terms like 'natural'. A wide variety of conceptions or models of nature has been considered, many of them serving as correctives to each other: nature as organism or as mechanism; as mine, sink or sanctuary; as wild or as cultivable; as incomprehensible or as regular; as ownable or unownable; as natural capital or as a locus of independent value. Of these, some (such as the conceptions of nature as the wild counterpart of culture, and, more significantly, as the matrix of value) help to explain concern about threats to the whole terrestrial system of nature, as well as to local neighbourhoods, while others (such as the conceptions of nature as mechanism or as natural capital) suggest practices which could have alarming implications for the survival of nature's wildness and otherness. Being open to such a range of conceptions, nature should perhaps be regarded as possessed of an enigmatic quality. If so, the enigma extends to the natural environment too, both as local neighbourhood and as the global environment.

GLOBAL ENVIRONMENTAL PROBLEMS

As mentioned at the outset, people now alive need to tackle environmental problems such as global warming, the growing gaps in the ozone layer, the destruction of rainforests, the growth of deserts and the pollution of the oceans. We recognise these as problems intuitively, even though, as will be argued, a value-theory is needed before problems can properly be identified as such. (A possible explanation is that these would be problems for *any* at all plausible value-theory.) They are also environmental problems, concerning as they do transactions between humanity and the natural environment; sadly, there are many other problems of this kind. Also they are global problems, in one or another sense of 'global'.

What makes a problem global? Not all environmental problems are global ones: lead pollution from ancient mine-workings, for example, or localised plagues of locusts. At least two senses of 'global' are relevant. Some problems are global because of an accumulation of local problems of the same kind, such as the slicks from numerous oil-tankers, traces of which are now spread around all the seas and oceans; another example is acid rain, which defoliates forests downwind from many particular chimneys, but which is reaching global proportions, so widespread is its distribution. These, then, are globally recurrent and cumulative problems. Other problems, however, are global because of the interconnectedness of global systems such as the global carbon, nitrogen and water systems and the global weather system. Thus emissions of carbon gases, in whatever quantities, are producing significant meteorological impacts not only regionally (as in Indonesia and neighbouring states in 1997–8) but also globally on ocean currents, wind patterns and ice-caps, and are almost certainly causing a rise in sea-levels worldwide. Another example is radioactive strontium, generated by tests of nuclear weapons of the 1950s and early 1960s, and now distributed globally in the upper atmosphere, returning to earth wherever there is rainfall. Such problems are global in the distinct sense of being mediated by global systems, and are thus globally systemic problems.

Some problems are global in both these senses.[27] Thus carbon emissions from vehicle exhausts contribute to the cumulative problems of the pollution of centres of population worldwide, and also help to trigger systemic global climate change through the greenhouse effect. Meanwhile other, apparently local problems, such as the holes in the Antarctic and Arctic ozone layers, result from global systems which disseminate worldwide gases such as chlorofluorocarbons (CFCs), usually emitted in small quantities in places far distant from the polar regions. This example shows how apparently small and localised actions can affect the global environment, whether those who perform them are aware of this impact or not. An ethic is clearly needed for the coming millennium that takes such global environmental impacts into account.

Not all global problems are predominantly environmental problems; inflation and poverty are examples to the contrary, despite their environmental consequences, although those concerned about the global environment would be unwise to neglect either. (The relation of poverty in particular to environmental issues will be discussed in Chapter 7 below.) The problems resulting from globalisation form a further example, problems which are centrally economic and cultural. Nor are all environmental problems global ones, as we have just seen; indeed it is often threats to a locality (plans, for example, to fell some familiar trees) which generate environmental concern. Yet those who delve into local problems often discover wider, and sometimes global, implications or parallels.

On any account, global environmental problems are plentiful. Cumulative problems plausibly include, in addition to those already mentioned, noise pollution, traffic congestion (which most obviously affects the built environment, but indirectly impacts on the natural environment too), deforestation (which also contributes to global systemic problems), losses of coral reefs, extinctions of species (often impacting on local or regional ecosystems), and losses of fertile soil, whether through erosion, salination or other chemical pollution (each of them more or less global processes). Meanwhile further systemic problems relating to climate change include desertification, freak weather such as hurricanes, and the loss to cultivation of previously fertile regions, while among systemic problems relating to economic and social systems can be included shortages of fresh water, industrial pollution, water-borne diseases, destruction of watersheds and of habitats such as rainforests, losses of non-renewable resources and the various problems stemming from poverty and from rapid population growth.

While social and economic systems are in theory alterable through human decisions, global systemic problems relating to natural systems are often incurable by these means, and are also often more acute than cumulative problems because the systems concerned are the only ones of their kind (or, if there is life on other galaxies, the only ones to which we have access); repair through transplants from another such system is out of the question. Doubtless global systems are robust enough to recover from stressful episodes; their survival of the eruptions of Krakatoa and of Mt St. Helens are often mentioned in this connection. But unless the Gaia hypothesis is correct in claiming that global systems have inbuilt perennial self-restorative capacities[28] (and here that hypothesis, if incorrect, offers its most dangerous and misleading counsel), there may well be no way of curing or even repairing severely overstressed global systems. If so, it becomes all the more important to recognise what is at stake, and to prevent natural global systems being subverted or put seriously in jeopardy.

A human-centred value-theory might find less of a problem in, for example, some amount of habitat loss than most broader value-theories would, but even on this narrow basis there are limits to what is tolerable. Besides, while some of these problems generate beneficial side-effects (vineyards in the Thames Valley, for example, resulting from climate change), and are thus not problems from every perspective, yet they all remain global problems of one sort or another from the perspective of almost any value-theory whatever. At the same time, most of them result, at least in part, from acts of a trivial or routine kind on the part of ordinary people in the course of their daily routine, whether cooking food, earning their living or travelling to and fro. Granted the widespread and far-reaching impacts of our actions, we can scarcely avoid taking seriously the need for some kind of ethic tending to limit the worst of these impacts, and to foster concern to conserve rather than destroy whatever is of value, and generally to promote non-destructive ways of life. As Hans Küng and Karl-Josef Kuschel put it in their *Global Ethic*, we cannot but 'condemn the abuses of Earth's ecosystems'.[29] Admittedly the burden of responsibility falls in part on governments, corporations and other organisations, but awareness of this presupposes that an ethic is needed which applies to their actions too.

There is, then, an interconnected worldwide or global system of nature, within which actions of individuals or groups in one place are liable to generate repercussions anywhere (and in some cases everywhere) in the system. This system transcends and encompasses all people now alive. In some senses it makes the difference between life and death for us all (or, for the unborn, between life and non-existence), and certainly every prospect for a worthwhile life for every person and for all creatures depends on it. This is the global environment. For the same reasons, the concept of the global environment is indispensable. Indeed every intentional environment, every local environment, every landscape and every neighbourhood depends on the global environment, and could not exist without it.

Hence claims that commitment and loyalty to the global environment are impossible are themselves confused. People who recognise the dependence of themselves and their surroundings on something greater have in every age shown themselves capable of loyalty (and sometimes of devotion) to whatever was believed to be this greater being, whether Nature, the Earth Mother, the gods or the Creator; hence there is nothing impossible in commitment to the cause of the global environment, based on belief in a comparable dependence, and in the shared destiny and common interest of creatures sharing this dependence. Thus the widespread tendency to speak of loyalty either to the planet or to its biotic community does not always betoken an overgeneralised rootlessness, but may express a clear-headed commitment to the shared system of systems, appropriate to an age of global interdependence and of global threats to the survival of humanity, to the quality of life of our successors and to the continued existence of most of Earth's species.

NATIONAL AND INTERNATIONAL RESPONSIBILITIES

'Think globally, act locally',[30] runs the environmentalist slogan. This is a salutary saying, as far as it goes. The need to think globally has been amply borne out already; and the widespread and recurrent need for action to protect local environments and neighbourhoods, and to galvanise support from neighbours and other local people, is also self-evident.

But to stop here would be a counsel of despair. For many environmental (and related) problems are, as we have seen, global problems. They affect global systems, and they are also propagated by global systems, whether natural or social, or, in other cases, they accumulate globally and thus achieve worldwide proportions. Here, local loyalties are completely insufficient. They are often necessary, as in their absence awareness of the need for common efforts and concerted action might lack for a spring-board. Yet if people's loyalties stay confined to the local level, they will remain ludicrously inappropriate to these challenges, like mice pitted against mountains. Unless individual agents and communities in all countries (or at least in a majority) act in solidarity and mutual support, many of the problems will remain unsolved.

Some problems could be tackled by single nation-states, or by coalitions of states; others cannot be tackled without the agreement of all or at any rate most countries. Where a particular country could avert an environmental problem without causing equally grave or more serious problems, there must be a strong case for the conclusion that it has a responsibility to do so; no smaller or lower-level body is likely to have the necessary power, and no overarching body to have the necessary authority.

Often, however, transnational corporations have greater power and resources than 'sovereign' states. Frequently, this puts them in a position analogous to that of middle-ranking countries, for they alone often have power over issues such as the continuation or discontinuation of the cultivation of wetlands or the felling of forests. While the economic system may make them answerable only to their shareholders, the external costs (or 'externalities') of their operations may well affect people in all continents and for generations to come, and other species too. Thus companies trading in armaments have impacts on Third World environments through the massive diversion of resources which their activities generate. International bodies such as the World Bank and the International Monetary Fund have even greater power and impacts, though with different chains of answerability. Such wide powers bring matching moral responsibilities.

There are also cases where nothing short of international agreement, involving concerted action in all territories, and regulation governing all companies of whatever size, is proportionate to the problems. The emission of CFCs appears to be one such problem, and global warming another; and these examples are unlikely to prove the only ones. It might seem that nothing can be said about responsibilities in this sort of case until first ethics and then the ethics of international relations have been discussed. But where the future of humanity and all sentient life on Earth is at stake, and remedies are to be had at this and no other level, it can reliably be predicted that strong international responsibilities at this level will be found to exist. Any account of ethics or of international relations with implications to the contrary attests its own bankruptcy. Related responsibilities will exist, if so, on the part of individuals to promote the kinds of international agreement and policies which are necessary.[31]

Some may be sceptical about all talk of a global ethic, suspecting here support for the encroaching tentacles of globalisation, or a hidden desire to impose a homogeneous global code, blind to cultural diversity and potentially hostile to minority peoples. Suffice it to reply that the preservation of diversity, both biological and cultural, and the protection of minority rights are likely to figure prominently in such an ethic[32], and will figure in the ethic presented in later chapters, and that far from promoting globalisation such an ethic could be wielded in criticism of the arrogance of great powers, multinationals and the global organisations (like the International Monetary Fund), which are often its standard-bearers. Far from having to foster uniformity, homogeneity, or other drab impositions, a global ethic can nourish self-determination, and the emancipation of localities from centralising bureaucracy, at the same time as fostering co-operation based on shared destinies and love of a shared environment.

The concept of the global environment, then, is far from an illusion, or from a substitute for properly rooted local loyalties. As we have seen, the shared environment which it denotes underpins these loyalties, and itself demands matching loyalties. More than that, it also gives rise to far-reaching responsibilities at national, transnational and international levels. At some times and places, these global responsibilities may go virtually unheeded; yet on them as well as on traditionally recognised individual responsibilities the survival of all Earth's communities, human and non-human, now depends.

NOTES

1. Kenneth Mellanby, 'Environment', in Alan Bullock and Oliver Stalybrass (eds), *The Fontana Dictionary of Modern Thought*, page (henceforth p.) 207a.
2. Mellanby, 'Environment', p. 207a, and 'Ecosystem', p. 190b.
3. David E. Cooper, 'The Idea of Environment', in David E. Cooper and Joy A. Palmer (eds), *The Environment in Question*..
4. Tim Ingold, 'Beyond Anthropocentrism and Ecocentrism', unpublished presentation to a Workshop on 'Ethics, Economics and Environmental Management' of the Swedish Collegium for Advanced Study in the Social Sciences, Uppsala, 1995, p. 4.
5. Ibid., pp. 2–7.
6. Ibid., p. 17.
7. Cooper, 'The Idea of Environment', pp. 179–80.
8. Nigel Dower, 'The Idea of the Environment', in Robin Attfield and Andrew Belsey (eds), *Philosophy and the Natural Environment*, pp. 146–7 and 151–2.

9. Andrew Belsey, 'Chaos and Order, Environment and Anarchy', in Attfield and Belsey (eds), *Philosophy and the Natural Environment*, pp. 162–3.
10. Dower, 'The Idea of the Environment', pp. 145–6.
11. Ibid., pp. 152–4.
12. Cooper, 'The Idea of Environment', p. 178.
13. Quoted in Ernest Hemingway, *For Whom the Bell Tolls*, p. 3.
14. Dower, 'The Idea of the Environment', pp. 148f.; he attributes the quotation from Piet Hein to Frank Barnaby, *The Gaia Peace Atlas*, p. 192.
15. Belsey, 'Chaos and Order', pp. 162–3.
16. Dower, 'The Idea of the Environment', pp. 149–50.
17. Bill McKibben, *The End of Nature*.
18. Neil Evernden, *The Social Creation of Nature*.
19. Holmes Rolston, 'Nature for Real: Is Nature a Social Construct?', in Timothy Chappell (ed.), *The Philosophy of the Environment*, 38–64
20. James E. Lovelock, *Gaia: A New Look at Life on Earth*; *The Ages of Gaia: A Biography of Our Living Earth*.
21. Henry David Thoreau, *Walden*, p. 280.
22. Iris Murdoch, *The Sovereignty of Good*, p. 84.
23. Gerard Manley Hopkins, 'The Windhover', *Poems and Prose of Gerard Manley Hopkins*, p. 30.
24. Thoreau, *Walden*, p. 279.
25. Are Nylund et al., *The Commercial Ark*, p. 21.
26. Karl Marx, *Capital*, vol. 3, p. 776.
27. This distinction derives from B. L. Turner II et al., 'Two Types of Global Environmental Change', *Global Environmental Change*, 1(1), December 1990, 15–17.
28. Lovelock, *The Ages of Gaia*.
29. Hans Küng and Karl-Josef Kuschel (eds), *A Global Ethic: The Declaration of the Parliament of the World's Religions*.
30. From René Dubos. See Gerard Piel (ed.), *The World of René Dubos*, Part 8, 'Think Globally, Act Locally: Local Solutions to Global Problems'.
31. L. Jonathan Cohen, *The Principles of World Citizenship*, p. 88.
32. As they do in the Rio 'Earth Summit' Declaration on Environment and Development. See Wesley Granberg-Michaelson, *Redeeming the Creation: The Rio Earth Summit: Challenges for the Churches*, pp. 86–90, Principle 22.

CHAPTER 2

GLOBAL ETHICS AND ENVIRONMENTAL ETHICS

INTRODUCTION

Without some kind of ethic (a theory of right and responsibility) and some kind of axiology (or value-theory), we lack guidance and direction for tackling problems, whether global, environmental or otherwise. What is more, we even lack a satisfactory basis for identifying problems in the first place. In this chapter some alternative understandings of ethics and of value-theory will be considered, and a substantive ethical stance will be presented and defended.

One basic issue in value-theory concerns the range of things that matter. In this connection there is a strong case for revising the traditional view that only human beings and their values and interests matter, or traditional anthropocentrism. For it is difficult to credit that nothing but our own species matters, morally speaking, just as it is difficult to credit that what matters is just our own family and friends, or just our own country, and nothing else besides. The same difficulties confront the variety of anthropocentrism which holds that absolutely everything exists for the sake of humanity, and for it alone (metaphysical anthropocentrism); as the philosopher Descartes pointed out, the view that all the galaxies exist simply for our sake is ridiculous.[1]

The alternatives to anthropocentrism include sentientism, which accords moral recognition to all creatures with feelings, and thus capable of pleasure and suffering, and only to such creatures; biocentrism, which recognises the moral standing of all living creatures; and ecocentrism, which regards ecosystems and the biosphere as having moral significance independent of that of their members. Abandoning anthropocentrism is likely to involve moving in one of these directions. True, there is also a case for remaining anthropocentrist, if what is meant by this is that we cannot help making all our valuations with human faculties and from a human perspective;

but this entirely sensible and harmless 'perspectival anthropocentrism' (as Frederick Ferré calls it)[2] is far removed from traditional anthropocentrism (the position just mentioned), and gives it no shred of support. On the contrary, people who recognise simply that animal suffering matters independently of human interests are already committed to discarding that kind of anthropocentrism.

A CHALLENGE TO THEORIES OF ENVIRONMENTAL ETHICS

But before considering what kind of stance should be adopted instead, we should reflect on Tim Ingold's challenge to all these 'centrisms'. All of them, he suggests (including even centrisms which do not make humanity central), distort our relation to the environment, establishing a boundary between humanity and nature. All of them thus ignore our pre-ethical engagement (or commitment) to our environments, and proceed as if we were detached observers, located outside the environment (whether understood as laboratory, mine or sanctuary), and as if we were capable of standing back from our unavoidable involvement with a whole network of relationships both with human beings and with non-human components of the environment. He suggests that it is from such relationships that any ethical system grows. Our own (equally unavoidable) part in these relationships consists in 'the kind of sensitivity and responsiveness which is the natural counterpart of [this] close and intimate involvement'; and for this pre-ethical stance, ethical codes are no substitute.[3]

Ingold does well to remind us that people have commitments before they begin thinking about ethics; as he implies, most people already have a range of related motivations by this stage (and far from merely egoistic ones at that), and hence such motivations do not have to be charmed into precarious existence out of an apathetic void. Yet to characterise people's pre-ethical commitments in ethical terms such as 'sensitivity' and 'responsiveness' imports far too much specific content into the diverse range of people's actual feelings about their environments. Moderately insensitive people can still have some kind of engagement with their native or adoptive environment, and of course have their responsibilities every bit as much as sensitive ones; while the sensitivities of quite sensitive people often turn out on reflection to be misinformed or misdirected, involving, for instance, an exaggerated or aggressive local patriotism. Pre-ethical engagements are no substitute for ethics.

These reflections have a bearing on the suggestion that recognition of pre-ethical commitments undermines ethical debate, such as the debate between anthropocentrism, ecocentrism and biocentrism, or shows it to be grounded in a confusion. For while pre-ethical commitments may make ethical commitments more feasible, they have not reached the stage of being ethical commitments, and they do not even begin to supersede the importance of ethical reflection. Most people would agree that we need to reflect on future generations, and on issues such as how like or unlike ourselves they may be expected to be, if we are to discover what sensitive individuals or agencies should do in their regard. In just the same way, we need to reflect on distant peoples, and also on non-human species, before we can get them too into ethical perspective. Ethical reflection can appeal to and take account of widespread intuitive judgements, as above over the significance of animal suffering, and thus need not operate as if human beings led lives detached from one another and from their environment; but it is far from exhausted by this appeal, and we should be immensely impoverished if it were so.

COSMOPOLITANISM AND COMMUNITARIANISM

In this book (and the other works in this series), a universalist ethical stance known as 'cosmopolitanism' is defended. Like the other authors in the series, I maintain that ethical responsibilities apply everywhere and to all moral agents capable of shouldering them, and not only to members of one or another tradition or community; and that factors which provide reasons for action for any agent, whether individual or corporate, provide reasons for like action for any other agent who is similarly placed, whatever their community may be or believe. The stance of cosmopolitanism is sometimes thought to disregard the networks of relationships in which moral agents find themselves; and these communities or networks are sometimes considered to generate limits to ethical norms and principles, as well as to motivation for compliance with them. Based as it is on community boundaries, this position is known as communitarianism, a stance opposed to cosmopolitanism. (Incidentally, cosmopolitanism and communitarianism are each consistent with anthropocentrism, biocentrism and ecocentrism alike, except where communitarians claim to have found ethical boundaries relevant to the scope of morality. So the different debates between them can be treated separately, up to the point where they are shown to intersect.)

Communitarianism begins with a creditable belief in the value of community, and with the defensible premise that moralities mostly arise from, and are learned within, communities. Sometimes it simply comprises 'a form of social criticism that is aimed at the disappearance of community in modern society'.[4] But it usually proceeds to the dubious conclusions that all moral rights and obligations depend on relationships, and that where there are no ties of community there are no moral obligations, nor moral motivations either. Sometimes it is held, partly for these reasons, that whatever is locally accepted is right, and should be recognised as such. Some communitarians even hold that external appraisals of the morality of a national, ethnic or local community make no sense, since meaningful discourse itself depends on shared community membership.

Cosmopolitans, believing as they do in universal moral responsibilities, can recognise the value of community, the moral inspiration often provided by communities, and the desirability of an upbringing among people of shared values. But they need not accept that agents bereft of relationships (like the protagonist in John Fowles' novel *The Magus*)[5] have no responsibilities. Nor need they hold that we have no responsibilities with regard to people (or other creatures) who themselves lack relationships. More importantly, they need not accept that there are no obligations between communities or across community boundaries. Not even people unaware of having any remaining family or community ties are free to treat others however they please, or to perform random acts of vandalism. Nor, happily, need cosmopolitans (nor the discerning reader) accept that the possibility of moral motivation ceases for those who have no community, let alone *towards* such unfortunates, nor that whatever is locally accepted is thereby invariably right, nor that moral discourse between members of disparate communities need be empty or at cross-purposes, as if travellers or journalists or tourists could never find common moral ground with the people of other cultures that they visit. Sometimes it will be so, but why must it always be so?

Communitarians sometimes accuse their critics of rootlessness. As mentioned in Chapter 1, David Cooper writes of global environmentalists that 'At home everywhere, today's intellectual is at home nowhere in particular. It would be no surprise if his idea of an environment would be The Environment.'[6] Nigel Dower has compared the tone of this remark to Alasdair MacIntyre's criticism of cosmopolitanism 'that in making people citizens of everywhere it makes them rootless citizens of nowhere.'[7] Now the criticisms both of MacIntyre and of Cooper are often on target, in MacIntyre's case exposing the lack of content in certain forms of liberalism, and in Cooper's case, holistic advocacy (for example, on the part of Deep

Ecologists) of a sense of oneness with the natural environment, or the planet as a whole, as if the subject's separate identity could be merged in the environment and discarded thereby.

But these criticisms often take an overgeneralised form. As Brenda Almond has pointed out in reply to MacIntyre, liberalism sometimes constitutes a substantial tradition with values which (she implicitly suggests) world citizens can share;[8] whatever view you take of liberalism, you would be ill-advised to assert that liberalism does not form a tradition, or that it is always neutral about values. Cooper also exaggerates. On the one hand, many people are members of multiple communities, without losing their roots in the community of their birthplace or upbringing; I am fortunate enough to be one of them. On the other hand, refugees and displaced persons often form new and vibrant communities, in the spirit of Aviezer Tucker's remark that 'The natural home of humanity is the dry land of the planet.'[9] Often (come to that), so do groups campaigning for justice for the world's poor, or for wilderness or wildlife preservation; having wide concerns does not make people rootless. 'Cultivating an unreflective familiarity with an environment' (if this really can be cultivated, as Cooper suggests)[10] is not the same as belonging or having roots. In any case, unreflectiveness, as Andrew Belsey remarks, is not a characteristic of human beings or of their relation to their local environment.[11] Nor is an unreflective familiarity necessary for membership in a community. Community membership can often be deliberately acquired or cultivated, occasionally through the foundation of a new community.

So cosmopolitans need not be rootless; and mobility is consistent with belonging to one or more communities and environments, and also with having a sense of responsibility, whether narrow or wide. For a sense of responsibility need not be confined to particular communities and their members, or to particular environments, much as local loyalties and a sense of place can nourish and renew a person's sense of responsibility. Besides, an increasing number of people, particularly those who trade or travel internationally or belong to international bodies (such as churches), are aware of being 'members of the global human community',[12] with its shared problems and possibilities (which are discussed further in Chapter 11). But before cosmopolitanism and the form of it upheld in this book are expounded further, its historical origins and some key stages in its development should be briefly reviewed.

KEY STAGES IN THE HISTORY OF INTERNATIONAL ETHICS

The possibility of ethical relations between different human societies has been apparent at least since the time of Augustine (AD 354–430), who in *The City of God* stressed the unity of the human species and also that of God's purposes for humanity.[13] Earlier still, Stoics such as Epictetus (c. 55–135 AD) had commended regarding oneself as a citizen of the world, among a person's other roles;[14] but with Augustine some of the ethical implications of intersocietal ethics were elicited, for example in his attempt to distinguish between just and unjust wars. Later, Thomas Aquinas (1224–74) supplemented this body of just-war theory, focusing, however, on the relations between different Christian societies.[15]

The issue of relations with non-Christian societies came forcefully to prominence with the Spanish conquest of Central America. In this connection, Francisco de Vitoria (c. 1483–1546) argued, on the basis of Aquinas' teaching, that it was unjust to make war on a people just because of their unbelief, or because they did not accept Christianity when it was first proclaimed to them; however, violence was justified to prevent practices like cannibalism, which were against the law of nature, and demonstrably wrong both to Indians and Europeans.[16] Similarly, his contemporary, Bartolomé de Las Casas (1474–1566), argued that it was justified to make war to prevent human sacrifices, understandable as such sacrifices might possibly be, but only if the number of lives saved was likely to exceed the number of the casualties of war (a notable appeal to the balance of consequences).[17] Both Vitoria and Las Casas believed in principles which applied to the actions of Europeans and Indians too, as well as principles governing how Christians should behave towards non-believers. Las Casas in particular believed that, while the distinctive features of Indian culture must be appreciated, the common human nature and reasoning capacities of people of all cultures make them subject to common responsibilities nonetheless.

A key further step was taken by Alberico Gentili (1552–1608) and Hugo Grotius (1583–1645). In contrast with their scholastic predecessors, both these theorists sought to found international relations on a secular basis, independent of theology. 'Let theologians keep silence about matters outside their province!', remarked Gentili, who grounded international law in universal agreement, whether explicit or tacit.[18] For his part, Grotius appealed to human nature, and to the law of nature which he believed to be implicit in it, as valid even if it were to be conceded that there is no God.[19] Thus belief in the difference between just and unjust wars, and thus in the global scope of ethics, was argued by both theorists to be independent of religious beliefs. Whether or not we regard their particular appeals to nature and to human nature as ground-breaking or persuasive, international ethics here importantly acquired once more, as among the ancient Stoics, a secular form and basis, capable of acceptance by people of any religious commitment or of none. The view that ethical reasons apply to everyone, irrespective of religious beliefs, just as prudential reasons do, is clearly crucial to a global ethic.

Also arguing on a secular basis, Immanuel Kant (1724–1804) grounded international morality on the Categorical Imperatives (maxims capable of being universally adopted) which he considered that any rational being must recognise. Kant regarded states as persons from the moral point of view, for which it would be rational, without abandonment of sovereignty, to enter into a confederation of free states for the sake of peace and mutual protection. In support of his claims, he draws attention to the verbal homage which all states pay to morality by claiming to have justice on their side. Likewise, morality demands related conduct of individuals, and any moral agent is censured who adopts a maxim which, if universalised, would make lasting peace impossible. In other words, it is a duty of every human being to work towards a confederation of free states, or cosmopolitan society.[20] Morality thus applies to individuals and states alike, and to the relations between states and between societies, and it is not impossible for these relations to evolve into an ethical international order. Further, these claims themselves form examples of cross-cultural moral truth,[21] belief in which Kantian cosmopolitanism upholds. (Appropriately, Kant himself used the term 'cosmopolitan', in the title of his essay 'Idea for a Universal History with a Cosmopolitan Purpose'.)[22]

Some alternative varieties of cosmopolitanism will be discussed in the coming section, but it is worth mentioning here that aspects of Kantian cosmopolitanism have been recognised as a serious contribution by some contemporary theorists, alongside the *Realpolitik* of Machiavellianism and the Grotian belief in an international order of nations. Thus Hedley Bull, developing Martin Wight's characterisations of these theories of international relations,[23] writes of a cosmopolitan concept of justice which seeks to derive rights and duties neither from national loyalties nor from international law, but from the promotion of the common good of humanity, a concept which he regards as of special relevance to ecological or environmental issues, although he is pessimistic about the prospects of its realisation.[24] This appeal to the common good clearly incorporates some non-Kantian, but still cosmopolitan, features; some non-Kantian varieties of cosmopolitanism will shortly be discussed.

But one objection to cosmopolitanism can be set aside without delay, the charge of Eurocentrism. Although Wight's theory of international relations has been accused of this form of ethnocentrism,[25] and although cosmopolitanism in particular has sometimes been associated with Enlightenment beliefs such as the belief in perpetual progress,[26] these associations are inessential to cosmopolitanism, and the kind of cosmopolitanism depicted by Bull and based on the common good of humanity need have nothing Eurocentric about it at all.

CONSEQUENTIALISM AND OTHER KINDS OF COSMOPOLITANISM

Rather than appealing, like Kant, to universalisable maxims, whether or not there is any chance of their being universally accepted, the position to which I adhere (and which I have defended elsewhere)[27] justifies practices by the balance of foreseeable consequences for all the parties affected (as compared with the consequences of alternatives), and justifies single actions on this same basis in cases where no such practice yet applies. Individuals and bodies with this stance take responsibility for shaping the future, not least through taking into account unintended but foreseeable consequences (regarded by Kant as morally irrelevant). Disregarding unintended outcomes is potentially disastrous, particularly where global systems are at risk. In this regard, I am closer to Las Casas than to Kant, and also to utilitarians, who standardly advocate promoting the balance of happiness over unhappiness. But consequentialists need not restrict their account of value and disvalue to human happiness and unhappiness, nor even to human interests; as indicated above, such an anthropocentric value-theory is excessively narrow, and consequentialists have no need to follow mainstream utilitarianism (at odds here with its founder, Jeremy Bentham) in adopting such a position. In that way they can take seriously foreseeable impacts of action which would affect non-human creatures of the present or the future.

Furthermore, consequentialists also hold that like interests count alike, wherever these interests are situated, and are thus committed to rejecting both the view that relationships must be present before obligations can arise, and the view that obligations extend only towards community members. They therefore endorse the views of Stoics and Christians that morality extends to all human beings, the belief of just-war theorists that it applies to the relations between societies and between states, the view of Vitoria and Las Casas that this includes peoples not sharing one's own religion, and that of Gentili and Grotius that it does not depend on religious beliefs being held at all. They are also free to agree with Kant that states have an obligation to work towards a stable international order, and that individuals are also obligated to contribute to this process. While consequentialists have not always adopted these internationalist implications, the logic of their position implies that they should do so, or so I shall be arguing in later chapters.

Like Kantians, consequentialists can also recognise the possibility of cross-cultural moral truth (as claimed by moral objectivists), and the related possibilities of seeking and finding it (affirmed by moral cognitivists). While all consequentialists accept that morality applies across cultural boundaries, independent arguments are needed to uphold these additional claims about the status of moral discourse and beliefs. This is not the place fully to rehearse these arguments (which I have presented elsewhere).[28]

However, the types of available argument may be briefly mentioned. One concerns making sense of our talk of knowing the difference between right and wrong, and also of moral awareness, and the presuppositions which accompany such talk. Another concerns the way in which alternative theories, which seek to relativise morality to the attitudes of particular groups or communities, reduce moral claims to sociological ones, and do not account for the fact that moral claims and principles present interpersonal reasons for action. Another points to the possibility of agreement concerning independent or intrinsic value, and thus concerning what form these interpersonal reasons sometimes take. A further argument recognises that there are many kinds of 'oughts', but adds that the range of reasons underpinning moral 'oughts' both limits and at the same time structures what counts as an 'ought' in such contexts, so much so that moral questions are decidable in principle, and admit of truth, just as prudential and technological questions are and do.

Besides consequentialism and Kantianism, a cosmopolitan approach could be combined with a contract theory or with a rights theory of society and of ethics. Cosmopolitan alliances are possible between consequentialists and theorists of either stamp. For example, negotiators might reasonably devise an international regime for carbon emission quotas on the contractarian basis of what would be agreed by all states in their own interests, irrespective of their bargaining position in the world as it is. However, particularly where environmental issues are in question, both of these theories suffer from serious shortcomings.

The appeal of contract theories partly depends on the assumption that whatever would be agreed to in a fair bargaining situation is just, and comprises a reasonable basis for social and possibly intersocietal co-operation. Thus they need not be concerned only with justice within particular societies, as John Rawls was in *A Theory of Justice*;[29] they can also be applied to relations between societies, as has been maintained by Brian Barry.[30] Some hold that they can even be applied to relations between successive generations, although the theoretical problems about bargaining between generations which are not contemporaries may well be insuperable.[31] Where, however, the assumption just mentioned clearly breaks down is over parties unable to bargain or make contracts which might safeguard their own interests; and of such parties, much the most numerically significant subset comprises all non-human creatures of the present and the foreseeable future. Since there could never be a contract which they could have agreed to without a change of nature and thus a loss of identity, we cannot assume that the rules derivable from imaginary ideal bargains, and such bargains only, are just and reasonable. If future interests and non-human interests are to be appropriately heeded, we need an ethic which neither excludes nor marginalises them, but which takes all these interests into consideration direct.

Here it may be suggested that we could instead argue from the rights of all parties liable to be affected by present actions and policies. I have no quarrel with taking rights seriously, wherever they can reliably be identified, although they should not, in my view, be regarded as fundamental in morality;[32] cosmopolitans of different persuasions, however, are free to unite in campaigning beneath the banner of human rights, including the right to a decent environment, and beneath that of animal rights too. The major problem for rights theories where the environment is in question, however, is that rights cannot be the basis of concern about that vast number of future people and other creatures whose identity is not yet determined. For such beings will not be individually better or worse off through present or imminent actions, as their very existence depends on these actions; different courses of action will result in different sets of future beings coming into existence,[33] and so no particular future beings (except those already conceived) can have rights against current agents. So we cannot reason from the rights of future individuals to ways of respecting these supposed future rights in the present. What we can do instead is to discover the responsibilities of current agents from the needs or alternatively from the likely interests of whoever there will be. This form of reasoning is acceptable to consequentialists, but cannot be based on rights.

A further problem for rights theories concerns non-human species. While most people accept that members of these species have a good of their own, only a limited range of them could at all plausibly have rights. The suggestion that all moral reasoning must have a basis in rights limits the scope of moral reasoning too narrowly. People who grant that whatever has a good of its own should be taken into consideration implicitly accept the view that, rather than base everything on rights and their bearers, we should recognise the broader class of the bearers of moral standing, or of moral considerability, entities whose good or interests should be considered or respected.[34]

Thus, however much we should go along with the advocates of rights, we need to supplement what they say when reasoning either about the future or about those non-human species (species of invertebrates and of plants included) which seem not to have rights. At least in these contexts we need to appeal to the good or the well-being of the relevant entities direct; and if this can be done, then either rights do not comprise the sole basis of morality, or (because this appeal to well-being could be made right across the board for all bearers of moral standing, whether they are bearers of rights or not) rights do not constitute the basis of morality at all. Accepting this is consistent with endorsing rights and the social or constitutional or international rules which enshrine them as of the utmost importance, albeit a derivative importance. The theory defended here endorses minority rights, and in general human rights and the rights of some animals too, on the basis just given, as was mentioned in Chapter 1.

Hence the position supported here is one which justifies rules, practices or (where these do not apply) actions on the basis of comparative balance of foreseeable consequences, as opposed to a basis of contracts, rights, or Kantian Categorical Imperatives. To this position, there are well-known objections, not all of which can be addressed in this book (and some of which I have attempted to address elsewhere);[35] some of these objections, such as those relating to our ignorance of the future, and the related impossibility of calculating distant costs and benefits, will receive attention in later chapters, when they become relevant. A further objection might comprise the view that consequentialism, in common with other forms of cosmopolitanism, proceeds as if a worldwide moral community were already recognised by all moral agents; if this were so, then consequentialism would collapse forthwith, as would cosmopolitanism. But this would beg the question against the view that some people have obligations to foster such a worldwide moral community, a view which presupposes that it does not yet fully exist. Clearly cosmopolitanism (and consequentialism in particular) can consistently uphold this view, and so the objection falls.

In any case, grounds have already been presented against the view that obligations can only exist within a community; and consequentialism, far from assuming that a worldwide moral community exists, does not even need to assume that such a community is fully possible (though I shall be arguing that it does involve obligations which presuppose that it is possible to move in the direction of such communality). In fact, consequentialism is applicable to situations of disaster and of minimal co-operation as well as to ones of near-Utopia, and to the range of cases in between; one of its strengths lies in its versatility in commending optimising strategies, however benign or adverse the circumstances and the prospects may be. Another dimension of its versatility lies in its capacity for fostering cultural toleration and respecting cultural diversity, except where agents are intolerant of toleration and diversity. Consequentialism is an ethic for all seasons.

CONSEQUENTIALISM AND PRINCIPLES OF ENVIRONMENTAL ETHICS

As we have seen, the objections to Kantianism, contractarianism and rights-theory have special force in the field of environmental ethics, and these are objections to which consequentialism is not subject. But it remains to be seen whether consequentialism can cope with environmental concern, or at least with the more consistent versions of that concern, and supply the basis of a satisfactory ethic for interaction with environments and with the global environment. My response to these questions is continued in the next chapter, where a stewardship or trusteeship stance is defended, and argued to be compatible in scope and values with the form of consequentialism defended here. In this chapter I am concerned not with an overarching metaphysical position, but with some relevant principles of normative ethics, and thus of value and obligation.

The question of moral standing (or considerability) needs to be addressed first. Here I claim that whatever has a good of its own (or would have, if brought into being) has moral standing. One ground for this view is that beneficence is central to morality, and that all such entities are capable of being benefited; another is that there is no other consistent stopping place when anthropocentrism is rejected, as it must be (see above); another is that ethical concern is possible for just this range of entities. These entities include all living creatures, both of the present and of the foreseeable future, and so my theory is a biocentric theory. Of course, if equal concern were to be advocated for all these creatures, the upshot would be an ethic by which it would not even be possible to live; an egalitarian form of biocentrism is indefensible and impossible. But moral standing is not moral significance, and bearing moral standing does not imply having equal significance with other such bearers. The moral significance of many creatures could be so slight as to be insignificant, except when the survival of large numbers (in the present or the future) is at issue, and even then their interests are outweighed when the vital interests of creatures with more sophisticated interests are at stake.

The moral significance of an interest depends on its intrinsic value, or its contribution to such value. Intrinsic value is understood here as a reason for action which is independent or non-derivative, and based solely in the nature of what has this value. All forms of consequentialism are committed to a theory of intrinsic value. For classical utilitarianism, for example, positive intrinsic value lay solely in happiness, and negative intrinsic value solely in unhappiness, and the same applies to its recent counterpart, Broad Utilitarianism.[36] For biocentric consequentialism, intrinsic value lies in the good or well-being of the bearers of moral standing. Following Aristotle, I take this good to consist in the development of the capacities essential to their kind, whether capacities for growth and reproduction (as in plants and animals alike), for mobility, perception and sentience (as in most animals), or for these plus capacities such as practical reason and autonomy, as in human beings. I also maintain that more complex and sophisticated capacities (such as that for autonomy) take precedence over simpler and less sophisticated ones, but only where both are at stake; no automatic priority belongs simply to membership of a sophisticated species, or simply to being human.[37] These claims would be resisted both by human chauvinists and by those who uphold the equality (on a one-for-one basis) of all living creatures; but neither of these parties can generate an ethic which is both consistent, tenable and defensible, unlike the position just introduced. (A more detailed defence of the value-theory presented here may be found in other essays of mine.)[38]

Many things also have value of other kinds, such as the value for observers of their beauty (inherent value), or their instrumental value to bearers of intrinsic value; and some of these, such as ecosystems, have so great an instrumental value, through facilitating the existence and flourishing of whole generations of creatures bearing intrinsic value, which could not exist without them, as to be capable of outweighing the value in the lives of even the individual human beings who could be brought into being and located there in their stead. In ways such as this, the biocentric theory which I am defending recognises the high value of many ecosystems, without recognising independent value in them, or moral standing either. Biodiversity could be regarded by biocentrists in the same light (see Chapter 8), being a precondition of all terrestrial value.[39] The continued existence of species likewise has a high instrumental value, because on this depends the very possibility of the existence of every possible future member of the species in question, together with the value which they would have; accepting this does not involve recognising intrinsic value for species, unless, like their members, they can be shown to have a good of their own. But this would have to be a good independent of that of their actual members and of their possible members, and it is implausible that their good is completely independent in these terms.

Because the theory defended here rejects claims about the moral standing and the intrinsic value of species, ecosystems and biodiversity, it is to be distinguished from ecocentric theories, which affirm such claims. But where ecocentric theories also recognise the moral standing of individual creatures and the intrinsic value of their flourishing, they thus cover the moral standing and intrinsic value of these creatures twice over (at least), once as individuals and once as members either of ecosystems or of species, or possibly as both; and this seems an unnecessary duplication. And once biodiversity is recognised as a precondition of all terrestrial value, and valuable for this reason, no point remains in assigning it an additional value on a par with the individual creatures which it nourishes.

Consequentialism goes on to put forward related criteria of right, wrong and obligation. These criteria concern not the maxims of actions and whether they can be universalised, as in Kantianism, nor whether deeds comply with rules which everyone might agree to, as in contractarianism, nor whether they uphold natural rights, as in rights-theory, but with the foreseeable difference made either by practices or (where no practices apply) by individual actions or policies to good and bad states of affairs, that is, to intrinsic value and disvalue. When practices, policies or actions are being considered or appraised, their foreseeable impacts on all affected parties (that is, all affected bearers of moral standing) are to be taken into account. This is a global ethic in several dimensions, since the impacts to be considered are restricted neither temporally nor spatially, and since they supply reasons for action or restraint for all agents, in whatever community they may be situated.

The global scope of this ethic makes it relevant to practices affecting distant places, such as the world trading system, with its pronounced environmental impacts for the environment of Third World countries, and for practices affecting the distant future, including the generation and storage of nuclear wastes. Unintended impacts of (say) five hundred years hence are ignored by Kantianism, and have no secure place in rights-theory or contractarianism; yet where they are avoidable, those who knowingly produce them have a clear moral responsibility in the matter, and it is one of the strengths of consequentialism that it recognises and underlines this. This and other implications of a consequentialist ethic will be investigated in greater detail in the coming chapters. Because consequentialism allows us to appraise technological and social systems, it also has the resources to go beyond 'end-of-pipe' (clean-up) solutions, and to advocate interventionist initiatives (such as programmes to enhance the efficiency of energy generation and use) where processes and outcomes would foreseeably be optimised in terms of overall value.

Meanwhile the related value-theory allows us to identify as environmental problems not only the causes of widespread costs for humanity (such as traffic pollution) but also damage or harm to non-human species, their ecosystems and their habitats, for the sake of non-human kinds. Biodiversity loss (which includes loss of natural habitats) now turns out to be a loss not only for humanity, present and future, but also for the creatures concerned, and for their successors whose existence is precluded. Similarly there are multiple reasons for identifying as global problems the destruction of forests, of wetlands and of coral reefs, reasons relating to their non-human inhabitants as well as their human beneficiaries. And while these are examples of global problems of the cumulative kind, the value-theory also underpins recognition of global environmental problems of the systemic kind (see Chapter 1), including global warming and gaps in the ozone layer. Such systemic problems involve costs for most if not all species; at the same time they are capable of distorting the evolutionary process itself, by changing the conditions in which it operates, and thus the range of creatures which it can generate. While cosmic processes are not intrinsically valuable, the theory recognises their overarching value by recognising the value in the lives of existing species (including their current populations) and also the value in the lives of the possible species which they can facilitate or, if distorted, can pre-empt.

While further implications of the ethical theory just presented will be elicited in Chapters 4 to 11, the role of humanity as trustees of nature, protecting the evolutionary process and its diverse products, will be further discussed in the next chapter.

NOTES

1. René Descartes, *Principles of Philosophy,* Part III, Principle 3; *Philosophical Works of Descartes*, vol. I, p. 271
2. Frederick Ferré, 'Personalistic Organicism: Paradox or Paradigm', in Robin Attfield and Andrew Belsey (eds), *Philosophy and the Natural Environment*, p. 72.
3. Tim Ingold, 'Beyond Anthropocentrism and Ecocentrism', unpublished presentation to a Workshop on 'Ethics, Economics and Environmental Management' of the Swedish Collegium for Advanced Study in the Social Sciences, Uppsala, 1995, pp. 11–16.
4. John O'Neill, 'Should Communitarians be Nationalists?', *Journal of Applied Philosophy*, 11, 1994, 135–43, p. 141.
5. John Fowles, *The Magus.*
6. David Cooper, 'The Idea of Environment', in David E. Cooper and Joy A. Palmer (eds), *The Environment in Question*, pp. 171–2.
7. Nigel Dower, 'The Idea of the Environment', in Attfield and Belsey, *Philosophy and the Natural Environment*, p. 144.
8. Brenda Almond, 'Alasdair MacIntyre: the Virtue of Tradition', *Journal of Applied Philosophy*, 7, 1990, 102–3.
9. Aviezer Tucker, 'In Search of Home', *Journal of Applied Philosophy*, 11, 1994, 181–7, p. 186.
10. David Cooper, 'The Idea of Environment', p. 163.
11. Andrew Belsey, 'Chaos and Order, Environment and Anarchy', in Attfield and Belsey, *Philosophy and the Natural Environment*, p. 163.
12. Jonathan Glover, 'The Research Programme of Development Ethics', in Martha Nussbaum and Jonathan Glover (eds), *Women, Culture and Development: A Study of Human Capabilities*, at p. 138.
13. Robert Nisbet, *History of the Idea of Progress*, pp. 60 and 287.
14. Epictetus, 'Discourses', 2.10ff., in A. A. Long and D. Sedley (eds), *The Hellenistic Philosophers*, vol. I, p. 364.
15. See extracts from Augustine and Aquinas, in Evan Luard (ed.), *Basic Texts in International Relations*, 28–32

16. Francisco de Vitoria, 'On the Indians', in Evan Luard (ed.), *Basic Texts in International Relations*, 145–9.
17. Bartolomé de Las Casas, *In Defense of the Indians*; cited in Vittorio Hösle, 'The Third World as a Philosophical Problem', *Social Research*, 59(2), 1992, 227–62, at pp. 239–40.
18. M. G. Forsyth et al. (eds), *The Theory of International Relations*, 15–24.
19. Grotius, *On the Law of War and Peace*, Prolegomena, 9–11, quoted in Forsyth, *Theory of International Relations*, pp. 44–5.
20. Immanuel Kant, 'Perpetual Peace: A Philosophical Sketch', in Hans Reiss (ed.), *Kant: Political Writings*, 93–130; Thomas Donaldson, 'Kant's Global Rationalism', in Terry Nardin and David R. Mapel (eds.), *Traditions of International Ethics,* pp. 145–6.
21. Donaldson, 'Kant's Global Rationalism', in Nardin and Mapel, *Traditions of International Ethics,* p. 143.
22. Reiss, *Kant*, 41–53.
23. Martin Wight, *International Theory: The Three Traditions.*
24. Hedley Bull, *The Anarchical Society*, 2nd edn 1995, pp. 80–7.
25. Hedley Bull, 'Martin Wight and the Theory of International Relations', in Wight, *International Theory*, ix-xxiii, at p. xxii.
26. As in Stephen Toulmin, *Cosmopolis, The Hidden Agenda of Modernity.*
27. Robin Attfield, *Value, Obligation and Meta-Ethics.*
28. Ibid., chs 3 and 12–14.
29. John Rawls, *A Theory of Justice.*
30. Brian Barry, *The Liberal Theory of Justice.*
31. Attfield, *Value, Obligation and Meta-Ethics*, pp. 14–17.
32. Ibid., pp. 7–8, 142–4.
33. Derek Parfit, *Reasons and Persons*, pt IV.
34. Kenneth Goodpaster, 'On Being Morally Considerable', *Journal of Philosophy*, 75, 1978, 308–25; Attfield, *Value, Obligation and Meta-Ethics*, ch. 2.
35. Attfield, *Value, Obligation and Meta-Ethics*, chs 7–11.
36. See Lincoln Allison, *Ecology and Utility.*
37. Attfield, *Value, Obligation and Meta-Ethics*, chs 4–6.
38. Robin Attfield,*The Ethics of Environmental Concern*, pp. 140–84, 204–9; *Environmental Philosophy*, pp. 203–20; *Value, Obligation and Meta-Ethics*, chs 4–6.
39. Paul M. Wood, 'Biodiversity as a Source of Biological Resources: A New Look at Biodiversity Values', *Environmental Values*, 6.3, 1997, 251–68.

CHAPTER 3

TRUSTEES OF THE PLANET

INTRODUCTION

There is nothing incoherent or absurd about loyalty to the planetary biosphere, or simply to our home, the planet Earth, and no need to be ashamed of such loyalty. Nationalists and communitarians are often right to feel pride in their own particular people, history and culture, but they have to recognise at the same time, as was argued in the last chapter, that ethics bestrides national and community boundaries, and that broader loyalties are both possible and, granted our global problems, often indispensable. Similarly, as was argued in Chapter 1, concern for local environments mandates concern for the shared, global environment, on which virtually all that we hold dear depends. In any case, no one can fence off their own climate from the global weather system and global climate change, as Robin Cook, the British Foreign Secretary, once remarked, any more than anyone can fence off their community from the rest of humanity. While identification with the biosphere is not commended here, and may actually be incoherent, we can hardly help, if we are consistent, caring for the preconditions of our being and of our identities, including the Earth, its biosphere, its systems and its biodiversity.

The kind of ethic equal to our situation extends far in time as well as space. As was seen in Chapter 2, our responsibilities concern future generations as well as our contemporaries. Sometimes they also extend to the past, in the sense of completing or continuing the projects of our predecessors. In short, it is inadequate to understand present agents as isolated individuals; without our ever having volunteered for membership, we seem to find ourselves involved as participants in a transgenerational community of moral agents, inheriting both benefits and burdens from our predecessors and passing them on to our successors. With regard to the environment, this suggests (if it is true) that we are entrusted by our forebears (whether or not they intended this) with the care of the planet and its systems, and that we perforce share this task with our successors, who will be among the beneficiaries if we play our part. In other words, we are trustees of the planet.

This makes it relevant to consider a particular tradition which recognises people as trustees of the planet, and which at the same time denies to the present generation (and sometimes to humanity as a whole) ownership of the land or of the Earth. This is the Judaic, Christian and Islamic tradition of stewardship. While environmental sensitivity does not of itself require adherence to this tradition, I shall consider here the relations between stewardship and environmental sensitivity, and the possibilities of a secular version of this tradition, to which non-believers would be free to adhere, before presenting some overall conclusions.

STEWARDSHIP AND ITS CRITICS

The ethic commended in the last chapter can derive support (or so I shall argue) from the tradition of stewardship, which has long been central to Judaism, Christianity and Islam. This support importantly means that adherents of these major religions can uphold this ethic through appeal to their own traditions; and if so, believers need not discard their religion in order to become environmentally sensitive in a consistent way (as is sometimes suggested). These religions have usually maintained that humanity is answerable to God, both for the use and for the care of nature, rather as the steward of an estate is answerable to its owner, or as trustees are answerable before the law for the goods which they hold on trust. They have also standardly maintained in consequence that our dealings with nature are subject to ethical constraints. Whatever our laws may say about property, another implication is that humans do not own the Earth, nor its lands nor its oceans, but hold or possess them on a provisional basis; hence their answerability.

However, the above claims about the stewardship tradition are controversial, since some writers represent the role (and the model) of stewardship as objectionably anthropocentric, managerial, aloof from nature, and thus no useful guide in environmental ethics, while others represent it as enlightened and heedful of nature's intrinsic value, but at the same time as unrepresentative of traditional religion.

While some of these criticisms may seem far-fetched, and not all the critics can possibly be right (for they clearly contradict one another), addressing the criticisms will elicit some important implications of stewardship. So I shall first discuss the ethical and political criticisms, and then turn to aspects of the historical development of stewardship beliefs, before considering the question of whether stewardship comprises a viable approach open to those who reject or no longer hold traditional religious beliefs.

One of the critics is Matthew Fox, a theologian who suggests that belief in stewardship represents God as an absentee landlord, and humans as serfs,[1] as if this belief deprived people of their freedom and spontaneity. That would be a serious defect, if it were true; but in fact neither tenant farmers nor trustees (who are both answerable to others, and thus analogous to the stewards of the biblical parables) are remotely as unfree or inhibited as this view suggests. So belief in stewardship does not abrogate freedom, or render the stewards serfs, or, come to that, make God an absentee landlord. Meanwhile belief in the answerability of human individuals and their communities conveys that we are ethically unfree in one significant way, unfree, that is, to treat the Earth just as we please; ethical limits to human transactions with nature are real and ought to be recognised. It is not these beliefs about answerability but their absence which contributes to ecological disasters.

Some related criticisms have been supplied by Clare Palmer. Her first suggestion is that the stewardship model separates God from the natural world, and makes respecting the world of nature less likely than a pantheistic or a panentheistic (or 'immanence') model would.[2] However, pantheism holds that the world is identical with God, and that God has no existence other than as the world; and this view precludes belief in the world being created by God, and respect being due to fellow creatures as God's creatures. So, quite apart from its being incompatible with the great theistic religions (Judaism, Christianity and Islam), it removes an important basis for respect for nature. Pan*en*theism, the belief that God is present or abides (or is immanent) in the world without being identical with it, is different, as it implicitly recognises God's transcendence, although, as Palmer recognises, it could lead to respect being focused not on nature but on the deity within it. But in any case the assumption behind the criticism should be questioned, namely that the stewardship model is incompatible with the immanence model; for it is far from obvious that the two models are mutually exclusive. Thus the stewardship view, as will be seen, is entirely compatible with belief in nature's independent value, a belief which has often fostered panentheism, as, for example, in the thought of Augustine.[3]

Palmer's second suggestion is that if the stewardship model is held in an absolutist manner (by which I take her to mean an uncompromising manner), this makes it difficult to accept the immanence model.[4] If this were so, the obvious solution would be to avoid absolutism about stewardship; for on any account it is implausible that stewardship encapsulates every facet of the relation between God, nature and humanity. But believers in stewardship need not in any case reject the belief that God indwells the world. For governments and owners (or generally those to whom stewards are answerable) typically live in the lands they rule or own, and so the stewardship model need not convey God's separateness; and if creation is continual (rather than a past event), divine activity might in any case be expected to pervade the natural order, rather than to pass it by. Nor need belief in divine immanence deter its adherents from using resources; sculptors, joiners and miners who become panentheists need not abandon their trades.

This discussion throws light on further criticisms from Fox, who declares: 'I reject the stewardship model (that God is an absentee landlord and we humans are serfs, running the garden for God); . . . We need mysticism – God IS the garden.'[5] For while love of nature may take the form of mysticism, which is sometimes inspired by panentheist beliefs, Fox's position is clearly pantheistic, and the pantheist attempt to identify God with nature, as we have seen, stipulates that there is no creation and no creator, and that the respect which might be due to fellow creatures has no religious underpinning as such; any respect for natural entities has to depend on some other basis (or on nothing). Later in this chapter, I shall return to the question of whether stewardship beliefs can be held by non-believers in God. But for present purposes it should be concluded that the suggestion that pantheism supplies a more adequate account of either religion or religious ethics or respect for nature than belief in creation and in stewardship is unconvincing.

STEWARDSHIP: A FULLER STATEMENT

Before turning to political, economic and historical criticisms of stewardship, it is appropriate to consider a more detailed expression of such beliefs.

> We all share and depend on the same world, with its finite and often non-renewable resources. Christians believe that this world belongs to God by creation, redemption and sustenance, and that he has entrusted it to humankind, made in his image and responsible to him; we are in the position of stewards, tenants, curators, trustees or guardians, whether or not we acknowledge this responsibility.
>
> Stewardship implies caring management, not selfish exploitation; it involves a concern for both present and future as well as self, and a recognition that the

> world we manage has an interest in its own survival and wellbeing independent of its value to us.
>
> Good stewardship requires justice, truthfulness, sensitivity, and compassion. It has implications . . . for individuals, organisations, and states.[6]

This statement from the General Synod Board for Social Responsibility of the Church of England involves claims that go well beyond the scope of this book, although its support for the concept of a global environment and for a global environmental ethic, and its acceptance of the independent value of non-human interests are significant, cohering, as they do, with positions defended here. Of greater immediate relevance is the range of metaphors used for the trust believed to be held by humans from God; it shows that belief in stewardship need not cast humanity specifically in the role of an ancient task-master of slaves or that of a medieval bailiff set over serfs, not least because the contemporary metaphors of curators and of guardians are at least equally in place as that of ancient or medieval estate-manager. The belief that the world belongs not to humanity but to God will also be seen to be significant.

POLITICAL CRITICISMS

Yet the charge continues to be made that stewardship presupposes a hierarchical social order of control and obedience, symbolising and inadvertently teaching despotism rather than democracy.[7] Undeniably the metaphor of stewardship in matters of resources derives from the teaching of Jesus, who deliberately drew on contemporary social life in his parables, and thus unavoidably it reflects, in part, the hierarchical social structure of the Roman Empire, as well as the more socially radical traditions of Judaism. But none of this has prevented the teaching of Jesus taking root in societies with very different forms of social order, including democratic and (sometimes) egalitarian ones. Such societies, like hierarchical ones, need to appoint officers specifically charged with the care of nature and the use of natural resources, and have proved able to interpret the stewardship model accordingly.

So the issue concerns not the origins of the metaphor of stewardship, but its current message; need it convey despotism, or at least unrepresentative social arrangements? If so, it would clearly be ill-fitted for coping with the global environmental problems of a new millennium. But depicting humanity as in a position of trust with respect to nature does not involve understanding society or government as either undemocratic or unrepresentative; if anything it commends democratic debate, so that the members of society can jointly discover or decide how to exercise their role. No relations of domination within humanity receive the least support from the stewardship model.

Trustees, however, are subordinate to the authority to which they are answerable, which, in the case of traditional stewardship beliefs, is God. Is this kind of subordination or answerability objectionable? Certainly if God is not believed to exist there would have to be some other form of answerability (see below), or else none at all. But if there is a creator, the suggestion that humans are not God's creatures and subjects cannot arise, unless being God's subject is equated with subordination to some hierarchical form of society or to some human bearer of divine authority (and here it is the supplementary beliefs about hierarchy or divine authority which need to be contested). Belief in humanity as stewards, however, implies not a need for social hierarchy but answerability (as opposed to ownership) with regard to the natural world; this belief makes neither humanity nor God a despot, but teaches a salutary humility, especially to people intent on remoulding the planet solely for human benefit.

Yet the related charge has sometimes been made that stewardship makes humanity a despot over nature. In particular, stewardship has been designated as an 'anthropocentric ethic', capable of advocating interference with the entire surface of the planet to enhance nature's productivity.[8] The charge of despotism is paradoxical, as the classification of traditional attitudes to nature of John Passmore contrasts despotic attitudes and stewardly ones;[9] yet if stewardship authorises changing the face of the entire planet in the interests of (say) productivity, then it could be considered not only anthropocentric but despotic too, and Passmore's definitive contrast between despotism and stewardship would be annulled.

This objection is partly an economic one, and would apply by extension to secular versions of stewardship beliefs if it fitted religious ones in the first place. But it clearly does not fit any religious versions at all like that of the General Synod, which recognises the independent value of the natural world, thus rejecting instrumentalism, and urges caring management as opposed to selfish exploitation. While caring management is undeniably a form of management, it is not managerialism, which implies a preoccupation with instrumental aims and values to the exclusion of all others. Caring management also implies recognition of constraints on instrumental approaches, and concern for future needs as well as present interests,[10] and thus, effectively, some amount of letting-be for both species and ecosystems; or, in other words, some amount of forbearance from management.

The objection is also implicitly historical, and certainly some advocates of stewardship, such as Jean Calvin, have held an anthropocentric view of creation;[11] yet even Calvin held that the beasts were to be treated with respect and not misused, but nourished and cared for, being creatures of God.[12] While these beliefs recognise animals as mattering simply for the sake of the creator, they are far removed from managerialism, and are close to the view that animals matter because of what they are. Indeed other advocates of stewardship, such as John Ray, Thomas Tryon and Alexander Pope, were soon adopting a more biocentric view of non-human creatures;[13] and this supplies further strong evidence against stewardship being essentially anthropocentric.

The same applies to the tone of the historical figure who first explicitly applied the language of stewardship to the natural world, Sir Matthew Hale. For Hale, 'Man' is God's 'Steward, . . . Bailiff, or Farmer of this goodly Farm of the lower World', the justifications of whose authority include 'to preserve the face of the Earth in beauty, usefulness and fruitfulness'. While humanity is to enjoy the fruits of nature, people are also to preserve species (and improve them), and, in addition to all this, to prevent the destruction of natural beauty.[14] There is no suggestion here that beauty is to be preserved just for the sake of humanity, and thus Hale's position seems to have been neither managerial nor anthropocentric. True, the mandate of humanity includes 'to limit the fiercer animals' rather than to respect their wild habitats. Nevertheless the charge of managerialism is off-target for Hale and his successors, let alone the charge of supporting planet-wide interference with nature in the human interest.

Stewardship can thus be defended against the above range of ethical and political objections. The distinct criticism that it does not express the position of the Bible and is unrepresentative of most of the Christian centuries will be discussed in the next section, which concerns history.

STEWARDSHIP IN HISTORY

I want to acknowledge at the outset that, implicit as the message of stewardship had earlier been in the parables of Jesus, explicit advocacy of the stewardship of possessions and resources began with Calvin in the sixteenth century, and that its first direct application to animals, plants and the rest of nature can be credited to Hale in the seventeenth century. However, this does not mean that stewardship beliefs both about resources and about nature were not substantially present in the Bible and in the intervening centuries, albeit in different language; the key components of these beliefs have roots much deeper than the early modern period.

Thus in Genesis 2, Adam is told to dress and keep the garden, apparently a role both of productive use and of conservation or protection. Palmer's suggestion that the contents of the garden and the various animals are represented as created solely for the sake of the gardener[15] conflicts both with the mandate to keep the garden (which suggests it has some kind of value of its own) and also with the other creation narrative (Genesis 1). In this narrative everything that has been created is seen by God to have been 'very good', and that not only for human purposes, as the plants are given as food to all the animals.

Elsewhere the Earth is understood as belonging to God (Psalm 24), and the land as not owned by humanity but as a leasehold (Leviticus 25:23), and as held subject to ethical requirements concerning the support of the poor (Leviticus 25:25–55; Deuteronomy 15: 1–11). Hence the passages about human dominion (Genesis 1:26–31; Psalm 8) have to be understood as concerning a conditional tenancy, and not unconditional domination. As Palmer stresses,[16] passages like Job 38–41 imply that the animals were not made for humanity, and that the wilderness is sufficient to itself and needs no gardener. Yet none of this suggests that the authors of Job would reject human responsibility with regard to the beasts (recognised elsewhere in the corpus of wisdom literature at Proverbs 12:10) or for letting the wilderness remain intact for them (as intended by the Creator, according to Psalm 104). While different passages have different emphases, the various strands still form a coherent picture.

The same applies to the New Testament. The teaching of Jesus about lilies and birds (Matthew 6:30; Luke 12:24, and so on) and about domestic animals (Matthew 12:11–12; Luke 13: 15–16; Luke 14:5) presupposes their independent value and moral standing, while his parables about stewards and accountability (Matthew 21:33–41; 24:45–51; 25:14–30 and the corresponding passages in Mark and Luke) concern not only the Church, as Passmore suggests, but implicitly the use and deployment of resources as well. Paul taught that terrestrial bodies of different kinds (humans, beasts, fishes, birds) have their own glory, comparable with that of celestial bodies (I Corinthians 15:39–41), and, like other New Testament writers (see, for example, John 1:1–14; Hebrews 1:2–3), he includes the whole created order in God's plan of salvation (Romans 8:19–22). Given also that Jesus and the New Testament writers took for granted Old

Testament teachings about creation, the land and the natural world, Passmore's view that there is little evidence for a stewardship interpretation of early Christian teaching cannot be accepted.[17] Stewardship is the clear message of the Old Testament, and consistent with the passages about human dominion there (Genesis 1; Psalm 8), as Eric Katz remarks, conveying also the standard interpretation adopted within Judaism,[18] while Clarence J. Glacken, in his masterly survey of historical attitudes to nature, readily interprets both Testaments in this sense.[19]

Stewardship, however, does not exhaust biblical or Christian approaches to nature. Passmore identifies as a distinct tradition the approach of co-operation with nature, in which the role of humanity is 'to perfect nature by co-operating with it',[20] but finds few if any traces of this tradition between pagan antiquity and the German metaphysics of the romantic period.[21] Yet the belief that the creation was deliberately left incomplete with a view to a challenge to human creativity and to scope for human improvements to nature pervaded the early centuries of Christian thought, from the time of Lactantius in the West[22] and Origen in the East[23] (both living in the third century AD), and was resuscitated in the seventeenth century by writers such as John Ray, William Derham and others.[24] Passmore is on surer ground when he writes of this and the stewardship tradition coalescing;[25] a good example is supplied by the Benedictine monasteries, which, throughout the period from Benedict (sixth century) to Bernard of Clairvaux (twelfth century), sought to enhance both the beauty and the fertility of their lands. These Benedictine attempts to improve the land also form a constructive example of stewardship, which René Dubos has aptly characterised as a paradigm of environmental responsibility. Granted the pervasive human need to derive food and shelter from the environment, as well as to conserve it, and the impossibility of preserving at all much of it untouched, Benedict is, for Dubos, a fitter patron saint of environmentalism than Francis of Assisi.[26] Similar attitudes to the enhancement and adornment of nature, combined with strong opposition to pantheism, were held by the fourth-century founder of Orthodox monasticism, Basil the Great,[27] and have persisted throughout succeeding centuries in the Eastern churches; his conception of humanity as partner of God in improving the Earth was popularised in the West by Ambrose, whose teaching probably influenced Benedict, among many others.[28] But this is not the place to discuss the history of attitudes to nature of the Christian centuries in further detail;[29] it is already clear that the stewardship tradition, often associated with ideas of co-operative improvement of the land, has been a central approach throughout these centuries, and not just a modern development.

A further strand among Christian attitudes to wild creatures is found in a New Testament passage about the forty days spent by Jesus in the wilderness: 'and he was with the wild beasts' (Mark 1:13). This sojourn among the desert animals is symbolically significant, because of longstanding expectations that the Messiah would make peace with the beasts. What is more, because this expression of acceptance and confraternity with wild creatures concerns neither conservation nor management, it supplements the message of stewardship. Richard Bauckham writes in this connection of 'peaceable companionship' with the wild creatures,[30] a posture which seems to have been consciously imitated by saints from St. Antony (third century) to the Celtic saints, to Cuthbert (seventh century) and to Francis of Assisi.[31] While no one has suggested that the generality of believers should also imitate Jesus in this respect, awareness of this aspect of the lives of the saints has kept alive a recognition that wild creatures deserve respect, and is likely to have been as influential as the largely anthropocentric teachings of theologians such as Aquinas[32] and Luther.[33] It has also served as a reminder that nature does not just consist in resources, and that stewardship is best understood in a non-anthropocentric sense. As Dubos says, 'Reverence for nature is compatible with willingness to accept responsibility for a creative stewardship of the earth.'[34]

Islam too has rediscovered ancient doctrines closely resembling the stewardship tradition of Judaism and Christianity. For Islam, the world belongs to God, and humanity is God's servant, Khalifah (caliph or vicegerent) and trustee of the Earth, accountable to God for its use and its care. The related responsibilities apply to all believers and all their activities, including all use of resources.[35] A relevant example of this teaching consists in the provision of Islamic law for 'himas', tracts of land set aside to remain undeveloped in perpetuity, of which thousands remain to this day.[36] While it is sometimes held that the world was created solely for human use, Al-Hafiz Masri maintains that according to Islamic law the natural elements are the common property of all creatures, and not only of human beings.[37] Thus the view is tenable that environmental problems in the Muslim world result from too ready an abandonment of Islamic insights and adoption of Western technology and beliefs about progress.[38]

Meanwhile another criticism of stewardship is relevant to the West of the post-Renaissance centuries, the period of capitalism. This is the criticism that stewardship is liable to ignore social justice, and might become 'reduced to a reasonable way of managing time, talent, and treasure', all, perhaps in the name of the kingdom of God; thus Mary Jegen.[39] Certainly if stewardship is reduced to this, or to the management of natural resources simply to maximise profits, then it falls short, and this has often happened in practice. But what it falls short of is the teaching of the Bible, and also of the medieval Church, for, as John Black points out, the teaching of Aquinas (thirteenth century) was that property beyond a man's necessity was owed, as of right, to the poor for their sustenance.[40] Such teaching was resuscitated and applied to the international stage by Pope Paul VI, who stated in *Populorum Progressio* that 'the superfluous wealth of rich countries should be placed at the service of poor nations . . . Otherwise their continued greed will certainly call down upon them the judgement of God and the wrath of the poor.'[41]

In general, while social justice and stewardship comprise independent commitments, twentieth-century Christian adherents of stewardship reject versions of stewardship unrelated to justice and to provision for the poor. Indeed Jegen's criticism does not condemn stewardship as such, but only reductionist versions.[42] I have elsewhere[43] defended early modern exponents of stewardship such as William Derham against related criticisms from William Coleman of too uncritical an endorsement of capitalist enterprise;[44] greater selectivity on Derham's part about contemporary commerce would have been in place, but longstanding Christian condemnations of greed and self-aggrandisement were never abandoned, and well cohere with stewardship. As with Islam, problems arise when ancient values are forgotten, rather than from remembering them.

One last charge against stewardship should be considered here, concerned as it is (in part) with history. For Palmer suggests that stewardship presupposes a pre-evolutionary view of nature,[45] and envisages humanity as set apart as God's manager on Earth. While this point may well require a revision of the position of some adherents of stewardship, such as the followers of Calvin, the view that everything was made for humanity conflicts, as we have seen, with the Bible, and with the stance of all who have rejected anthropocentrism. Where this view is rejected, humanity cannot be supposed to be called on to settle everywhere or to manage everything (and thus potentially to redeploy it), including the habitats of all the other creatures for whom the created order has (from this perspective) been made, despite Margaret Thatcher's reported claim that 'All we have is a life tenancy [sc. of the Earth] with a full repairing lease.'[46]

Palmer goes on to point out (rightly) that the idea that universal management is needed is a nonsense. Her point tallies with James Lovelock's remark that nothing worse could befall the planet than humanity becoming or trying to be stewards or managers of it.[47] But her conclusion can still be questioned: 'Stewardship is inappropriate for some of the planet some of the time, some of it for all of the time (the deep oceans) and all of it for some of the time – that is, before humanity evolved and after its extinction.'[48] For one thing, stewardship is not synonymous with interventionism, and is compatible with letting-be (appropriate for, say, Antarctica, an example of her own). There again, granted that there was no human responsibility before there were human beings, and there will be none after human extinction, responsibility remains possible for the entire sphere of nature which humans can affect (and not only for the sphere of human settlement or appropriation), and in the twentieth century this includes, for better or for worse, the deep oceans, the solar system, and much of outer space beyond it. Unless this extensive power is exercised with responsibility, global problems will be intensified. Thus the choice is between power exercised responsibly and power without responsibility. So, far from evolutionary theory making stewardship obsolete, twentieth-century technology actually makes an attitude akin to stewardship indispensable.

Whether such an attitude can be held in the absence of religious belief is a question which has now become urgent. It is tackled in the next section.

STEWARDSHIP WITHOUT GOD

Are beliefs in answerability possible where belief in God has disappeared? Belief in responsibility need not lapse in these circumstances; some acts and some omissions remain unacceptable, in view of their contexts and consequences, even if belief in the Kingdom of God is absent. Thus in his later philosophy, Martin Heidegger wrote of 'dwelling' (*das Wohnen*) with the things which comprise the natural environment; such heedful inhabitation implicitly involves the role of 'care-taker' (in the full sense of one who has '*Sorge*' or 'care').[49] Yet motives such as love and loyalty reinforce responsibility, and are prone to accompany answerability; so the question remains worth asking, because of the difference liable to be made to motivation if it can be answered affirmatively.

To whom, then, or before whom would secular stewards be responsible? In 1990 the Conservative government of Britain seemed to take the view that stewardship is an ethical responsibility, and 'an imperative' which must 'underline all our environmental policies', requiring us 'to look after our planet and to hand it on in good order to future generations'.[50] Admirable as was the global scope of this statement, its readers could be excused some unease about how far 'in good order' implies managerialism. Also, while it makes good sense to talk of obligations with regard to future generations, and such talk could comprise part of the basis of secular stewardship, we could not actually be answerable to generations which do not yet exist. Another possible answer is supplied by David Pearce and fellow authors, when in *Blueprint 2* they write that 'Humans should act as nature's stewards and conserve natural resources and the environment, for their own sakes and to preserve the interests of other creatures.' While the motivation which they applaud for conserving 'natural assets' is concern for human interests, they remark that this 'also conserves the environments of sentient non-humans and non-sentient beings'.[51] Despite differences at the level of theory, this position (which also explicitly rejects ecocentrism) in practice comes close to the consequentialist biocentrism commended in Chapter 2 above. Even so, the question about answerability remains unanswered, and this might be held to undermine talk of stewardship, except as a term for an ethic of this kind.

A less explicit but possibly more significant expression of stewardship forms the basis of the 1990 Code of the G7 nations, which speaks of 'stewardship of the living and non-living systems of the earth in order to maintain their sustainability for present and future, allowing development with forbearance and fairness'.[52] It may be that the anodyne nature of these words is what allowed the Economic Summit Nations (the G7) to adopt them. Yet their adoption commits the world's leading economic powers to efforts to tackle disruptions of natural systems, such as global warming, acid rain and ozone depletion. However, answerability to natural sytems is clearly out of the question.

More light is shed on secular stewardship from an unexpected quarter. For Karl Marx, while discussing the need to sustain the soil across the generations, wrote as follows about the impossibility of owning the Earth:

> From the standpoint of a higher economic form of society, private ownership of the globe by single individuals will appear quite as absurd as private ownership of one man over another. Even a whole society, a nation, or even all simultaneously existing societies together, are not the owners of the globe. They are only its possessors, its usufructuaries, and, like *boni patres familias*, they must hand it down to succeeding generations in an improved condition.[53]

In this striking (albeit anthropocentric) passage, 'possessors' can be translated as 'occupants', 'usufructuaries' as 'tenants', and the Latin phrase '*pater familias*' which Marx uses can be translated 'head and representative of household'. So his words convey that the current generation must bequeath the Earth to succeeding generations in the way that good representatives of family lineages hand down family resources. Marx seems to be saying not only (as Leviticus does) that the Earth cannot be owned, at least not by any one generation, but also that the reason why the current generation must bequeath it in an improved condition is that this generation comprises representatives of humanity, conceived as a transgenerational community.

While this passage can be held to show ecological awareness, it is as limited as the Conservative White Paper 'The Common Inheritance' of 1990 with regard to the value of nature; there is no awareness that the natural world consists in more than resources. Nevertheless, in accepting that people now alive do not own the Earth but hold it on trust for their successors, it comprises an early secular expression of stewardship; and it also evokes an answer to the problem of answerability through hinting that the current generation are answerable to the transgenerational community of humanity. Because this community, unlike future generations, has living members, it is not absurd to talk of answerability to such a community. Furthermore people's widespread sense of obligation to past members or to their memory makes this all the more credible, at least where living people are regarded as continuing the projects of the dead (see Chapter 4). Marx did not, of course, hold that responsibility or answerability attach equally to all current humans, since he was acutely aware of inequalities of resources and of power, arising from oppression. But his position still conveys the stewardship of humanity as a whole.

An alternative way of thinking about answerability focuses on the class of those who share (or have shared or will share) in the task of caring for the biosphere and the planet. For we may all be held to be answerable to the other members (or to the class as a whole) for our share in this task; if some default or defect, the greater becomes the burden of the others. Here the relevant class includes all moral agents: once again, a transgenerational community, consisting of all the individuals and organisations capable of responsible action and of making a difference to the world and its value. Humans now alive would, on this basis, be stewards and trustees of the planet, answerable to the ampler company of predecessors, contemporaries and successors combined;[54] and this approach has the advantage of making us responsible to agents entitled to complain if we shirk our part.

However, the scope of the community of moral agents is virtually co-extensive with that of the transgenerational community of humanity (except that moral agents would also include God). Hence little practical difference is made whichever of the two communities is invoked with relation to a secular theory. In either case, answerability remains a characteristic of stewardship, which turns out to be not simply an ethic, but capable of being understood as involving an appropriate metaphysical backdrop, and an implicit awareness of the company of those to whom our trusteeship is owed. Against this backdrop, obligations with regard to future generations come to appear more significant, since we, the current generation, are now seen to depend on the beneficiaries of our responsibilities for the continuation and in some cases completion of our tasks, just as we continue the projects and tasks of previous generations.

This is the place to consider Murray Bookchin's suggestion about the scope of the secular stewardship of humanity. Besides stressing the responsibility for the natural world conferred on human beings by evolution, and the distinctive attributes which facilitate this responsibility, Bookchin also claims that stewardship can take the form of intervention into natural processes, intervention as creative as the creativity of nature itself, of which it comprises a realisation.[55] To the extent that this means accepting responsibility for ecological impacts (including global environmental problems), and for taking steps to ameliorate or even cure them, this is a welcome suggestion. But if it involves either total management of the surface of the Earth, or attempts to redirect the evolutionary process, despite our abiding ignorance of its workings, it resembles rather a secular counterpart of the religious-based managerialism castigated by Palmer and others. Thus my verdict on the scope of stewardship consists in welcoming accounts which recognise a trusteeship extending to the impacts, actual and possible, of human action (and which thus cohere with the scope of the normative ethic of consequentialism), but rejecting accounts which (unlike biocentric consequentialism) represent humanity as authorised to act as if everything were made for itself, and as authorised to attempt to manage nature as a fiefdom rather than to conserve and care for it as a trust.

AFTERWORD

Effectively the same issue, of whether secular stewardship opens the way to exploitation of nature, is sometimes raised in the form of the suggestion that historically stewardship has 'desacralised' nature, authorising its investigation and its unlimited appropriation and use. Where 'desacralise' concerns rejection of the worship of nature, theistic religion has been a desacralising influence throughout its history, fostering worship of God alone, and (in recent centuries) the secular study of the natural world. But where 'desacralise nature' means representing nature as having no independent value of its own, neither theism nor stewardship implies anything of the kind. In the words of the General Synod, it can recognise that the world has 'an interest in its own survival and well-being independent of its value to us'[56] or rather that its living creatures do (as the books of Job and the Psalms attest). Both traditional religious stewardship and secular trusteeship imply that the sphere of our responsibility is as extensive as our powers, but that the scope of management has strong ethical limits. Some of the resulting dilemmas will be discussed in later chapters.

While most non-theistic religions (as well as theistic ones) include strands conducive to environmental sensitivity, and this makes possible the advocacy of an interreligious global ethic concerned in part with such sensitivity,[57] I have argued in this chapter that the stewardship tradition of theistic religions well equips them in particular to endorse and foster the kind of ethic commended here, and to offer reasons why it is important. I have also sought to show that secular versions of stewardship can be embraced without adherence to these religions, harnessed rather to a secular metaphysic and matching secular motivations. While the ethic defended in this book embodies reasons of its own (the intrinsic value promoted by its adoption and the intrinsic disvalue or evil averted), and requires neither a religious nor a secular metaphysic, nor related motivation, the availability of these underpinnings serves to enhance its overall credibility and attractiveness through the self-understanding conveyed of our role vis-à-vis the world of nature.

NOTES

1. Matthew Fox, lecture at St. James's Church, Piccadilly, London, 1990; cited in R. J. Berry, 'Creation and the Environment', *Science and Christian Belief*, 7(1), 1995, 21–43, p. 25.
2. Clare Palmer, 'Stewardship: A Case Study in Environmental Ethics', in J. Ball et al. (eds), *The Earth Beneath*, 67–86, p. 75
3. Clarence J. Glacken, *Traces on the Rhodian Shore: Nature and Culture in Western Thought from Ancient Times to the End of the Eighteenth Century*, pp. 196–202; H. Paul Santmire, *The Travail of Nature: The Ambiguous Ecological Promise of Christian Theology*, pp. 55–74.

4. Palmer, 'Stewardship'.
5. Fox, Lecture at St James's, Piccadilly, as quoted by Berry, p. 25.
6. General Synod Board for Social Responsibility, *Christians and the Environment*, p. 2 (summary).
7. Palmer, 'Stewardship', pp. 75–7.
8. Ibid. pp. 77–82; Richard and Val Routley, 'Human Chauvinism and Environmental Ethics', in Don Mannison et al. (eds), *Environmental Philosophy*, 96–189.
9. John Passmore, *Man's Responsibility for Nature*, pt One.
10. Thus Peter G. Brown, 'Toward an Economics of Stewardship: The Case of Climate', forthcoming in *Ecological Economics*.
11. H. Paul Santmire, *The Travail of Nature*, pp. 124f.
12. Keith Thomas, *Man and the Natural World: A History of the Modern Sensibility*, p. 154.
13. Ibid., pp. 155, 166–7; Peter Singer, *Animal Liberation: A New Ethics for Our Treatment of Animals*, p. 221.
14. Sir Matthew Hale, *The Primitive Origination of Mankind* (1677), quoted at John Black, *Man's Dominion: The Search for Ecological Responsibility*, pp. 56–7.
15. Palmer, 'Stewardship', p. 70.
16. Ibid.
17. Passmore, *Man's Responsibility*, p. 29.
18. Eric Katz, 'Judaism and the Ecological Crisis', in Mary Evelyn Tucker and John A. Grim (eds), *Worldviews and Ecology: Religion, Philosophy and Environment*, 55–70.
19. Glacken, *Traces on the Rhodian Shore*, p. 168.
20. Passmore, *Man's Responsibility*, p. 32.
21. Ibid., pp. 33–4.
22. Glacken, *Traces on the Rhodian Shore*, p. 181.
23. Ibid., p. 185.
24. Ibid., pp. 484, 423–4.
25. Passmore, *Man's Responsibility*, p. 32.
26. René Dubos, 'Franciscan Conservation and Benedictine Stewardship', in David Spring and Eileen Spring (eds), *Ecology and Religion in History*, 114–36.
27. Glacken, *Traces on the Rhodian Shore,* p. 192; D. S. Wallace-Hadrill, *The Greek Patristic View of Nature*, pp. 128–30.
28. Ibid., p. 196.
29. A brief overview is supplied in Robin Attfield, 'Christianity', in Dale Jamieson (ed.), *A Companion to Environmental Philosophy*.
30. Richard Bauckham, 'Jesus and the Wild Animals (Mark 1:13): A Christological Image for an Ecological Age', in J. B. Green and M. Turner (eds), *Jesus of Nazareth: Lord and Christ: Essays on the Historical Jesus and New Testament Christology*, 3–21.
31. Susan Power Bratton, 'The Original Desert Solitaire: Early Christian

Monasticism and Wilderness', *Environmental Ethics*, 10, 1988, 31–53; Helen Waddell, *Beasts and Saints.*

32. For a discussion of Aquinas' views, see Robin Attfield, *Environmental Philosophy: Principles and Prospects*, pp. 46–8.
33. Santmire, *The Travail of Nature*, pp. 124–5, 128–31.
34. Dubos, 'Franciscan Conservation and Benedictine Stewardship', p. 136.
35. M. Kamal Hassan, 'World-view Orientation and Ethics: A Muslim Perspective', forthcoming in Azizan H. Baharuddin, *Development, Ethics and Environment.*
36. Yassin Dutton, 'Natural Resources in Islam', in Fazlun Khalid and Joanne O'Brien (eds), *Islam and Ecology*, 51–67; see pp. 54–7.
37. Al-Hafiz B.A. Masri, 'Islam and Ecology', in Khalid and O'Brien, *Islam and Ecology*, 1–23, p. 6.
38. Lisa Wersal, 'Islam and Environmental Ethics: Tradition Responds to Contemporary Challenges', *Zygon*, 30, 1995, 451–9.
39. Mary Jegen, 'The Church's Role in Healing the Earth', in W. Granberg-Michaelson (ed.), *Tending the Garden*; cited at Berry, 'Creation and the Environment', 93–113, p. 26.
40. Black, *Man's Dominion:*, pp. 64–6.
41. Pope Paul VI, Encyclical Letter *Populorum Progressio* (on Fostering the Development of Peoples), paragraph 49.
42. Jegen, 'The Church's Role'.
43. Attfield, *Environmental Philosophy*, pp. 32–4.
44. William Coleman, 'Providence, Capitalism and Environmental Degradation: English Apologetics in an Era of Revolution', *Journal of the History of Ideas*, 37, 1976, 27–44.
45. Palmer, 'Stewardship', pp. 78–9.
46. Margaret Thatcher, address to Conservative Party Annual Conference, September 1988.
47. James Lovelock, talk on Radio 3, 10 June 1992.
48. Palmer, 'Stewardship', p. 79.
49. Bruce V. Foltz, 'On Heidegger and the Interpretation of Environmental Crisis', *Environmental Ethics*, 6, 1984, 323–38, pp. 336–7; Martin Heidegger, *Poetry, Language, Thought*, p. 47
50. Cmd.12200, *The Common Inheritance.*
51. David Pearce et al., *Blueprint 2: Greening the World Economy.*
52. Cited in Berry, 'Creation and the Environment', p. 34.
53. Karl Marx, *Capital*, vol. 3, p. 776
54. See Attfield, *Environmental Philosophy*, pp. 59–60.
55. Murray Bookchin, 'Thinking Ecologically: A Dialectical Approach', *Our Generation*, 18.2, 1987, 3–40.
56. General Synod Board for Social Responsibility, *Christians and the Environment*, p. 2 (summary).
57. Thus Hans Küng and Karl-Josef Kuschel (eds), *A Global Ethic: The Declaration of the Parliament of the World's Religions.*

CHAPTER 4

THE ETHICS OF EXTINCTION

INTRODUCTION

The extinction of humanity is now a significant possibility, not least through environmental impacts of human actions. Thus a nuclear winter could descend as the environmental impact of nuclear warfare, and this is just one of several conceivable catastrophes. Another is a pandemic resulting from biological warfare, or from terrorist action, or even from accidental spin-offs of experimentation. Chemical warfare could produce similar effects. Or imaginably, global warming could accelerate irretrievably; or the ozone layer could be lost, and the conditions of human life thus be undermined.[1] Another possibility, however remote, is a collision between our planet and an asteroid, of a kind which may account for the demise of the dinosaurs in the Cretaceous Period.

There is nothing inevitable about any of these threats; the point here, however, is their mere possibility, and the values which reflection on this possibility can disclose. Such threats cannot be entirely disregarded. For example, between stable geological periods the forces which underlie stability can produce sudden geological change, as at the end of ice ages;[2] again, in the right combination, small changes like the manufacture of chlorofluorocarbons (CFCs) can generate vast consequences, in this case imperilling the prospects of all life on Earth.[3] This particular threat may be under control, black markets permitting, but might not have been if CFC impacts on the ozone layer had not been discovered when they were, or if CFC emissions had begun a few decades earlier, before science was capable of studying their effects.[4] Besides, the very survival of humanity to date has happened against the odds; '[h]umans are here today', as Stephen J. Gould says, 'because our particular line never fractured – never once at any of the billion points that could have erased us from history'.[5] This granted, our future as a species is likely to remain a vulnerable one.[6]

While the probability of these various threats remains extremely low, the mere possibility of human extinction serves to remind us of values which we might otherwise ignore. Intuitively most people recognise that to cause or allow the avoidable extinction of humanity and other sentient species would be an evil of almost unsurpassable magnitude, and also that the knowledge that humanity was to be extinguished within the coming decades would make much of current life become futile and meaningless. Reflection suggests that the irretrievable loss of worthwhile life at some stage of the future would be an even worse disaster for humanity than actual extinction.[7] But while this is not seriously in prospect, except in the event of people discovering that extinction itself has somehow become imminent, human extinction itself still remains a serious possibility. However, the intuitions just mentioned about the morally disastrous nature of extinction are not easily explained. In this chapter, different explanations of these intuitions are discussed, together with the implications of discarding them.

PAST AND FUTURE GENERATIONS

The belief that avoidable extinction would be an evil might seem to be based on duties owed to future generations. Yet if those generations never live, apparently they can never be harmed, and thus nothing can be owed to them. Besides, the belief that the extinction of humanity makes much current activity meaningless seems to be based either on such duties to future generations, which apparently could prove empty and thus impossible, or on duties to past generations to ensure that their concerns and projects are continued into the future; but people who are dead apparently cannot be either harmed or benefited, and so these duties seem empty and impossible too.

Maybe these verdicts about duties to past and future generations are too hasty. Consider duties to the past. Remembering the dead, through anniversaries and memorials, is often held to be a responsibility of the living; and John O'Neill has cogently argued that the dead really can be harmed and benefited, depending on whether their projects languish or prosper.[8] A claim of David N. James serves to relate this conclusion to the possibility of human extinction:

> [I]f there is any obligation whatever to honour and respect those whose past activities have given us what we have, that respect would appear to rule out placing at risk the totality of their contribution.[9]

Yet the degree to which the dead could be harmed or dishonoured still appears insufficient to account for our intuitive responses to the possibility of the extinction of humanity. Imagine a situation in which the dead could only be honoured through special effort and at the cost of neglecting people on the brink of starvation, or in need of rescue from rising floods. Our responsibility would clearly lie where the greater difference to human well-being could be made, through intervention on behalf of the living, at the cost of neglecting the dead. But a still greater difference to human well-being than this would seem to be at stake where the continued existence of an entire human generation hung in the balance. While our duties to the dead would admittedly support, rather than clash with, human preservation, such duties carry no more than diminutive strength when measured against the sheer overwhelming imperative to preserve humanity.

THE PLACE OF FUTURE GENERATIONS

This reflection suggests a reconsideration of the implicit place of future generations in our values. But Jonathan Schell has suggested that the risk of human extinction during the coming decades carries a much more immediate threat, striking at the meaning of life itself. For what makes it possible for our lives to have meaning here and now is the existence of an ongoing, communal and intergenerational 'common world' (Hannah Arendt's phrase), and the imminent truncation of this common world actually makes life forego its meaning and in this way renders all our values valueless.[10] For Arendt, the common world is made up of human works of lasting value, and gives meaning to all other human activities;[11] the extinction of humanity, Schell implies, curtails both the duration of the value created, and the shared character of the activities which create it, thus undermining the value of these pivotal activities in the present, and thus of all human life henceforth.

James, who gives this argument sympathetic consideration, endorses John Cooper's claim that '[p]articipation in shared activities is a necessary condition of a fully meaningful life'.[12] This claim could be doubted. Perhaps solitary mystics lead a fully meaningful life, and the claim should be that participation in these activities is typically (but not invariably) such a necessary condition. And if a meaningful but not-quite-fully-meaningful life can be led in the absence of these activities, as the supporters of this claim might concede, maybe they should really just be claiming that shared activities enhance life's meaningfulness.

But James manages to show that there is more to the relation between shared activities and a fully meaningful life. (He has in mind activities like contributing to art or to an ongoing subject of study.) Shared activities are valuable for human beings because we are social beings. Shared activities, more than purely private activities, enable one to be continuously and happily engaged, and provide one with an immediate and continuing sense of the value of one's activity. The activity of others agrees with and confirms one's own activity, providing an antidote to flagging interest. Moreover, shared activities enhance the meaning of one's activity by putting it within a broader group activity, which itself becomes a valued end.[13]

While these are mainly empirical, psychological claims, they also go a long way towards showing that, normally and typically, shared activities occupy a central place in a meaningful life. Perhaps there are also conceptual ties between the capacity for these activities and a worthwhile human life; I have argued this elsewhere,[14] but my argument there concerned the intrinsic value of the exercise of each of a whole range of capacities, without turning on the shared or communal nature of the activities in question.

James' argument turns to activities stretching out continuously into the future. Such activities (for example, science, philosophy and the arts) can also be shared activities, at least in the sense that they are 'engaged in with the hope and anticipation that future generations will continue them or participate in them'.[15] It could be added here that people who engage in such activities often regard their activity as defined (in part) by the contributions of past generations, and see themselves as continuing a tradition. Such sequential collaboration increases the meaning and value of the activity just as concurrent participation does. This is again a perceptive psychological observation on James' part.

The argument continues as follows:

> Sequential shared activity bestows meaning and value on the agent's life here and now whenever these activities are engaged in with the hope and anticipation that future generations will continue and participate in them. Without this hope and anticipation, intergenerational activities would lose some or all of their point.[16]

But interruption diminishes the value of activities pursued for the sake of the ends produced; and we might reasonably add that it also diminishes the value of activities performed for their own sake where the context which makes them worth pursuing for their own sake includes the prospect of continuing, intergenerational participation. '[H]ope and anticipation of a future is a fundamental context or horizon for valuable and meaningful human activity.'[17] So the prospect of human extinction deprives these intergenerational activities of some of their meaning, and makes them comparatively valueless. Or so the conclusion would run. James takes this conclusion to apply both to activities like the construction of durable and lasting public goods such as buildings and bridges, of which the ends would be interrupted and nullified, and to activities like science, philosophy and the arts, of which the current intrinsic value might be held to turn, in part, on their character as involving the prospect of future participation. 'As Schell suggests . . ., only extinction can annihilate the prospect of future value.'[18]

This is how James relates his argument to nuclear omnicide and its implications:

> A nuclear holocaust does not merely promise to cause unimagined suffering and the death of those who are killed. Such an event would not only end our lives and prevent future generations from coming into existence; it would also render intergenerational activities and activities shared with contemporaries incomplete and futile – interrupting the career of humanity in midcourse and annihilating the value and meaning which otherwise would have existed.[19]

And writing a decade earlier, Edwin Delattre went even further:

> To the extent that men are purposive, . . . the destruction of the future is suicidal by virtue of its radical alteration of the significance and possibilities of the present. The meaning of the present depends on the vision of the future as well as the remembrance of the past. This is so in part because all projects require the future, and to foreclose projects is effectively to reduce the present to emptiness.[20]

Yet while all this shows that certain current activities would lose one of their central sources of value if life on Earth were shortly going to be obliterated, it does not show that life here and now would lose its meaning altogether. Even if we grant that participation in shared activities is a necessary condition of a fully meaningful life, the prospect of the curtailment of shared intergenerational activities would not spell the abandonment of shared activities in general. Would you, in these circumstances, give up (for example) conversation? Philosophy, science and the arts could, I suggest, also continue (in principle, right up to the last moment). So could sports like football; and it would be morally imperative that some shared activities should not be abandoned, such as the nursing of the dying. Given what was granted above about the value of shared activities, these shared activities could well continue to be sources of value, muted by gloomy anticipations, and beset, no doubt, by the reservations of some participants about whether these activities were really worth pursuing in the circumstances.

For there could be replies to such reservations. Reactions to news of the prospect of an early death can range from making peace with family and friends (hardly signifying loss of all sense of value) to refusing to be interrupted in (say) playing the piano, in completing a game of chess, or in sitting in the shade in one's garden. In part, the implication is that some activities, such as playing music or chess, are intrinsically worthwhile; and this would be one way to reply to the sceptical reservations just mentioned. Indeed this implication would supply sufficient reason to persevere with many current activities (some of them shared ones). While these activities could well be less pleasant than usual (pleasure usually depending on unimpeded circumstances), they would not for that reason have become meaningless.

Thus future generations may not play quite the place in our values maintained by James. However, his argument does not show nothing. It shows that a considerable source of belief in the worthwhileness of current activities is the prospect of continuing, intergenerational participation. This is well brought out when James considers the objection that not all value can depend on hope for the future, as some people are prepared to die for the sake of honour, and such self-sacrificial activity is profoundly meaningful. But, as James replies, the death of an individual is not comparable with the extinction of humanity. 'Self-sacrifice is a shared social activity. . . . The hero, the duelist and the martyr hope to be remembered as those who chose death before dishonor. Fully meaningful self-sacrificial activity is possible only where there is hope of a human future.'[21] And what applies to self-sacrifice holds of a wide range of much more mundane behaviour.

This granted, is it really credible that the sole or the central ground for preserving humanity is the role of future generations in enhancing the value of current activities? Imagine that all such activities were doomed, but humanity still could be preserved (perhaps by computers programmed at the right time to defreeze and then nourish human embryos in an underground environment which would become secure from fallout 500 years after a foreseen global nuclear holocaust). Can it seriously be held that, if no other route to human survival were available, there would still be no obligation and no reason to arrange for humanity to be preserved by this means? And to return from this fanciful example to the world as it is, can it seriously be maintained that the main or the only reason for making plans now for the energy needs of people likely to live when the world's current human population have all died depends on our current pursuits, and consists in the value which the present prospect of the activities of these future people confers on these pursuits?

WHAT IS WRONG WITH EXTINCTION?

Nevertheless, as John Leslie has remarked, many philosophers write as if there were no reason for preserving the human species beyond obligations either to the dead or to the living, and some as if there would be nothing wrong with allowing the species to extinguish itself, or even with actively extinguishing it ourselves, well before this would happen in the ordinary course of events.[22] Now the argument concerning the value of ongoing current activities already shows that the verdicts that there would be nothing wrong with allowing (let alone causing) premature extinction are unsupportable; for the prospect of premature human extinction deprives many (but not all) widespread current activities of their meaning and value. But, as has just been argued, there must be something else to explain the strength of the imperative not to allow or to make premature extinction come about, and to explain what it is that makes most people who contemplate the possibility of premature human extinction regard it as appalling. Cicero makes a parallel point: 'As we feel it wicked and inhuman for men to declare that they care not if when they themselves are dead the universal conflagration ensues, it is undoubtedly true that we are bound to study the interest of posterity also for its own sake.'[23]

Likewise the consequentialist ethic introduced and defended in Chapter 2 maintains that future people have moral standing (and future living creatures of other species too). Future generations have this standing even though their existence is contingent on current generations and the identity of future individuals is unknown at present; the good or ill of individuals who could be brought into existence count as reasons for or against actions or policies which would bring them into being. This in turn implies that where the existence beyond a certain date of individuals likely to lead happy, worthwhile or flourishing lives can be facilitated or prevented, there is an obligation not to prevent it, other things being equal. This does not mean that everyone should be continually having children; other things are seldom equal, and problems of human numbers mean that acting on this basis could easily produce overextended families, countries or regions, or an overpopulated planet, where extra people would spell misery for themselves and for the others (see Chapter 7). But it does mean that each life likely to be of positive quality comprises a reason for its own existence, and that countervailing reasons of matching strength (concerning the disvalue of adding this life) are required to neutralise such a reason.

There are many other implications, including the importance of planning for the needs of future generations (considered in later chapters). A further implication, more relevant here, is that humanity should not be allowed to become extinct, insofar as this is within human control, even if, foreseeably, a small minority of any given generation will lead lives of negative quality (lives which are either not positively worth living or actually worth not living), as long as, overall, the lives of that generation are of positive quality, and the positive intrinsic value of worthwhile lives outweighs the intrinsic disvalue of the lives of misery. Since each generation is highly likely to include some lives which are not worth living, however hard its members and their predecessors may try to raise the quality of these lives, this implication makes all the difference to the issue of whether causing or even allowing the extinction of humanity is a moral crime. People who think that preventing misery is always of the greatest importance have to take the view that human extinction should be tolerated or even advocated; but the consequentialist ethic defended here says otherwise. So, of course, say the widespread intuitions reviewed earlier.

A modified version of one of John Leslie's thought-experiments[24] could be used to test much the same issue. On each of numerous inhabitable planets, capable of supporting a large human population, whose members would predictably lead lives of positive quality, there will also be a person whose life will predictably and inevitably be of negative quality. For the purposes of the thought-experiment, these large human populations can be brought into existence by waving a magic wand. Should this be done? For consequentialists who believe in optimising the balance of intrinsic value over intrinsic disvalue, and in counting every actual and possible life as having moral standing, the answer is affirmative, even though the resulting population of each planet includes a life of negative quality. But theorists who prioritise the prevention of misery would have to hold that the answer depends entirely on whether the life of negative quality on each planet can be prevented; if it cannot, then none of these lives should be engendered. (Others too, including consequentialists, might also take this view if the addition of human lives were liable to harm the living creatures of these same planets; to make this thought-experiment a test case, we need to adopt the further assumption that no such harm would be done.)

This thought-experiment also has a bearing on human extinction. For the future of the Earth beyond a certain date (just after the death of the youngest person now alive) is in some ways similar to the situation of the planets just mentioned. The current generation could produce a population living then, most of them people with lives worth living, but only at the risk of producing a minority whose lives will foreseeably be miserable. If the happiness or the worthwhile lives of the majority do not count as reasons for generating those same lives, and hence nothing counts but the misery of the minority, or if the prevention of misery should be prioritised over all else, then allowing extinction is clearly mandatory, and so may be even genocide. However, as Leslie claims, the coexistence of hundreds of thousands of lives of positive quality with one life of misery is not morally disastrous, if the misery of the miserable life really cannot be alleviated.[25] (If of course this misery could be alleviated, whether by contemporaries or by the previous generation, then this might well be a morally disastrous situation, and alleviation would almost certainly be obligatory.) Consequentialism, then, does not mandate extinction, unlike several of the theories which stand opposed to it.

TOTAL USE, SELF-TRANSCENDENCE AND REASONS TO CARE

Consequentialism, however, is also sometimes accused of mandating either excessive population increases or the total appropriation of the surface of the planet by humanity, and there are people who would prefer a theory which would in some circumstances call for human extinction to a theory with these implications. But it has been explained above that consequentialism neither mandates nor encourages overpopulation; as I have argued elsewhere,[26] it advocates additions to the human population only if habitable spaces can be found (for example, on other planets) where such additions would neither harm other species nor lead to net losses of value. As for total use of the planet, consequentialists would find this objectionable partly because of the countless extinctions of species which would be implicated, together with the pre-empting of the value in the lives of all their future members, and partly because of the enormous impoverishment of humanity, which would forego the opportunity to experience wild places and to contrast the artificial environments of cities and countrysides alike with tracts of wilderness largely unaffected by human interventions. (These are substantially the grounds supplied in Chapter 3 against total management of the surface of the Earth as a supposedly acceptable version of stewardship.)

These consequentialist reasons against the total use of the planet also explain why consequentialists could not be expected to prefer the maximising of the human population at the expense of subsequent extinction. Jan Narveson suggests that if lives of positive quality should be maximised, then a scenario in which a larger human population is concentrated in a few generations, followed by extinction consequent on the exhaustion of resources, would have to be seen as preferable to a less large population spread out across a longer future.[27] But the interests of non-humans (dependent on the same resources), together with the interest of humans in the continued existence of non-human life into the indefinite future, require the other scenario (the one with a longer human future) to be preferred, because the balance of reasons support it and value across time is maximised thereby. Human suffering at the stage when resources are running out, plus demoralisation at the prospect of extinction, supplement the grounds for this conclusion.

Philosophers who raise the question of why the interests of future people matter, and thus comprise a reason for conserving either resources or natural systems, usually write as if this questioning is consistent with recognition of current interests and values. But this view confronts the problem of reconciling disregard for the interests of one period and of recognising exactly similar ones of another period, apparently a clear case of unjustifiable discrimination. One writer who has honestly faced up to the implications of writing off the future is Thomas H. Thompson, who recognises that the questions of 'Why care about future generations?' and 'Why be moral?' are in practice the same question.[28] According to Thompson, these questions were capable of affirmative answers for devotees of religious belief, and to some extent for those influenced by Enlightenment substitutes for such belief, such as belief in progress, but for contemporary secular people no reason is left for the preservation of humanity, whether 'forever' or at all.[29]

Ernest Partridge's response to Thompson's case is profound, albeit incomplete. People, he argues, have a psychological need to transcend their petty interests, and to identify with larger ideals, movements or causes; and caring for posterity is a central case of such self-transcendence. Self-transcendence typically involves love, and, as John Passmore has suggested, to love is to care about the future of what we love, and for its sake rather than for our own. Partridge bears out these claims by imagining that astronomers establish that events on the sun will extinguish all life and human culture from the face of the earth in two hundred years' time; such awareness would profoundly and enormously affect people now, because we need the future to lead fulfilled lives in the present.[30] This thought-experiment probably gains in forcefulness through not distinguishing between concern in the present for future generations of humanity and concern for those of other species; but there is no need to consider them separately here, as both are crucial to the value-theory of this book. The possible strategy (which technology might conceivably permit) of transporting both human culture and some of the species with which humans interact to another planet is not mentioned by Partridge; but its predictable appeal in the circumstances which he depicts further bears out his point about the importance of the future for present people.

Responses like Partridge's help to show how people are often motivated to care for individuals and groups beyond their own interests; and the issue of actual motivations has to be tackled in response to positions like Thompson's. But so does the issue of what we have reason to do, whether or not we are actually motivated; for the value of life and of quality of life (as opposed to their perceived value) cannot fluctuate with whether given agents care about it or not. And if life of a positive quality gives those agents who can promote or preserve it reasons to care, as morality presupposes, then this applies in principle to caring about future lives just as much as to present ones. Our intuitions about the appalling nature of allowing or actually causing human extinction point in the same direction. The ethic put forward in this book (see Chapters 2 and 3 above) recognises all this and turns out to supply reasons for caring, and not only guidance about policies and conduct.

If so, then anyone who accepts this value-theory and ethic, whether his or her outlook is secular or religious, implicitly recognises a multitude of reasons to care. Religious beliefs such as those discussed in Chapter 3 will sometimes accompany (and may underpin) this recognition; but belief in stewardship, as argued there, can adopt a secular form, not dependent on belief in perpetual progress, and equally capable of fostering this same recognition. Partridge's argument adds that self-transcendence, in which such caring is central, is also beneficial for the psychological health of those who care, averting the narcissistic self-preoccupation and alienation which is so prevalent in current society. Indeed he holds, reasonably enough, that people lacking self-transcendence should be pitied.[31]

So reflection on the possibility of human extinction serves to reveal a good deal both about our values and about our interests. It reveals both our concern for the continuation of shared human activities, and recognition of the positive value of every worthwhile life, whether present or future, thereby reinforcing the value-theory of the consequentialist ethic presented in Chapter 2 above, which dovetails to a nicety with the findings of the current chapter. But these values do not make provision for future generations a straightforward matter, important as it has been seen to be, nor solve how to reconcile it with provision for current interests and needs. In Part Two, issues concerning the conservation of resources and of ecosystems and concerning sustainable provision for both current and future needs will be addressed.

NOTES

1. See Roy Porter, 'The End is Nigh', *The Observer,* 14 April 1996.
2. Colin Tudge, *The Day Before Yesterday*, pp. 34–5.
3. Ibid., p. 74.
4. Ibid., p. 58.
5. Stephen J. Gould, *Eight Little Piggies*, p. 229.
6. Tudge, *The Day Before Yesterday*, pp. 361–3.
7. See Jan Narveson, 'On the Survival of Humankind', in Robert Elliot and Arran Gare (eds), *Environmental Philosophy*, 40–57, at p. 56.
8. John O'Neill, 'Future Generations: Present Harms', *Philosophy*, 68, 1993, 35–51; *Ecology, Policy and Politics*, ch. 3.
9. David N. James, 'Risking Extinction: An Axiological Analysis', *Research in Philosophy and Technology*, 11, 1991, p. 53; James ascribes this view to K. Kipnis.
10. Jonathan Schell, *The Fate of the Earth*, pp. 117–8.
11. Hannah Arendt, *The Human Condition*, pp. 49–58.
12. John Cooper, 'Aristotle on Friendship', in A. Oksenberg Rorty (ed.), *Essays on Aristotle's Ethics*, pp. 324–30; cited at James, 'Risking Extinction', p. 57.
13. James, 'Risking Extinction', p. 57.
14. Robin Attfield, *Value, Obligation and Meta-Ethics*, chs 4 and 5; *A Theory of Value and Obligation*, chs 3 and 4.
15. James, 'Risking Extinction', p. 58.
16. Ibid., p. 58.
17. Ibid., p. 62.
18. Ibid., p. 61.
19. Ibid., p. 62.

20. Edwin Delattre, 'Rights, Responsibilities and Future Persons', *Ethics*, 82, 1972, 254–8, at p. 256.
21. James, 'Risking Extinction', p. 59.
22. John Leslie, *The End of the World,* pp. 155–80.
23. Cicero, *De Finibus*, 3.64; cited in Stephen R.L. Clark, 'Environmental Ethics', in Peter Byrne and Leslie Houlden (eds), *Companion Encyclopedia of Theology*.
24. For some related thought-experiments, see Leslie, *End of the World*, pp. 181–3.
25. Ibid., p. 181.
26. Robin Attfield, *Value, Obligation and Meta-Ethics,* ch. 10; *Ethics of Environmental Concern*, ch.7.
27. Narveson, 'On the Survival of Humankind', pp. 41–42.
28. Thomas H. Thompson, 'Are We Obligated to Future Others?', in Ernest Partridge (ed.), *Responsibilities to Future Generations*, 195–202, p. 200.
29. Ibid., pp. 200–1.
30. Ernest Partridge, 'Why Care About the Future?', in Partridge (ed.), *Responsibilities to Future Generations*, pp. 203–20; John Passmore, *Man's Responsibility for Nature*, pp. 88–9.
31. Partridge, 'Why Care About the Future?', pp. 206, 214.

PART II
APPLICATIONS AND ISSUES

CHAPTER 5

GLOBAL RESOURCES AND CLIMATE CHANGE

In the Third World, maintains Anil Agarwal, 'environmental destruction is not an issue of quality of life but is a question of survival'.[1] The survival of humanity may not be at stake, but the survival of families and of communities often is, when wells or springs dry up, forests disappear, or the land turns to desert. Such processes can all be caused by human agency; in particular, consumption of resources in developed, Northern countries often does violence, albeit imperceptibly, to global weather systems and ecosystems, as when, in July 1998, Americans were officially requested to curtail carbon emissions to mitigate a heat-wave. As Alan Durning has written, 'Overconsumption by the world's fortunate is an environmental problem . . . Their surging exploitation of resources threatens to exhaust or unalterably disfigure forests, soils, water, air and climate.'[2]

In this chapter, current problems concerning environmental resources and sinks (nature's absorptive capacities, that is) and their use and conservation will be considered, in the light of climate change, and also of ethical principles which stress the importance of provision for the needs of all affected parties, present and future, human and non-human too. Application of the principles to the problems allows some solutions or ways forward to be reached and presented in this and the following chapters, subject to the perennial need for possible revision in the light of unforeseen or unforeseeable developments. But the concept of resources will first be explored.

THE CONCEPT OF RESOURCES

Natural resources are goods supplied by nature and available for consumption, use or enjoyment. They are usually divided into renewable resources, like land, fresh water, forests and fishstocks, and non-renewable resources, such as minerals and fossil fuels, subject to the proviso that bad management of fragile renewable resources (through excessive irrigation, deforestation or overfishing) can turn them into a non-renewable condition. Distinctions can also be drawn between resources which are consumed, such as fuel or food, and ones whose benefits consist not in consumption but in appreciation, such as landscapes, or in their continuing intactness and availability, such as soils and seas. While the depletion of nonrenewable resources is often considered a problem, the subversion of renewable resources, often through international transactions, is emerging as a greater one, at least where the interests of future generations are concerned.

But the classification of forests and fishstocks as resources raises issues which call for some qualifications to be made to seeing them in this light. If something is a resource, can it be treated as simply a means to our ends, or could it at the same time have moral standing and bear intrinsic value? Can resources also be liabilities? And are resources confined to goods conducive to human interests, or do the interests of other species come into the picture? Further, what are the scope and the limits of the concept of a resource? For example, is nature a resource? There again, are natural sinks, which are often contrasted with natural resources, to be counted among resources, as the above definition suggests? And can human beings and their skills be regarded as resources, as management theorists characteristically assume? These questions have an ethical importance, which must be investigated before any stance on the ethics of resource-use or resource-conservation can be reached.

To tackle first the question, 'Can resources also be liabilities?': such they most certainly can be. Take the example of CFCs and HCFCs (hydrochlorofluorocarbons). These substances were devised for their instrumental value as propellants and refrigerants, but have turned out to have disastrous side-effects for stratospheric ozone (which protects humans and other creatures from skin-cancer), and (as if that was not enough) also comprise one of the kinds of greenhouse gas now widely recognised to require curtailment to prevent excessive global warming. Fortunately implementation of the Montreal Protocol (agreed in 1987) and the subsequent Adjustments and Amendments are likely to prevent their manufacture and use (black markets permitting), and the ozone layer should recover within a century.[3] This example already shows that the fact that something can be put to profitable and beneficial use does not mean that it should be, and that the intactness of the global environment can depend on worldwide, concerted restraint from such use.

This lesson is underlined when the issue is considered of whether resources are to be regarded as simply means to our ends. Not even minerals can be treated in this way, if 'our' refers to the current generation, as future generations may need them; quite apart from the side-effects of using them, their depletion raises issues of sharing and of equity between generations, and the possibility that future people should be compensated for current use. Issues of equity also arise when resource-use in developed countries affects developing countries. Consequentialists are committed to taking all these issues into account; and to these issues we shall return.

However, many resources (as defined above) are also living creatures (such as cattle or fish) or systems of living creatures (such as forests or wetlands) or wild species (such as whales or elephants). But if each creature with a good of its own has moral standing (as I have argued in Chapter 2), then all these interests should be taken into account alongside our own, and carry some independent moral weight, albeit a varying weight in accordance with degrees of sophistication and of consciousness. And besides the intrinsic value of the good of living creatures, their populations and species are often important through their contribution to natural systems, systems indispensable for human cultures and also for non-human ways of life both in the present and for the foreseeable future. Nor should the aesthetic and symbolic values of species, landscapes and wilderness be forgotten. So the mere fact that something is a resource need not make our consuming it or even managing it as much as an option.

This is just as well, if we reflect that nature itself (incorporating, as it does, all mineral and biological resources) is sometimes considered as a resource. Yet, as the origin and matrix of all value among creatures, it is neither ownable, nor to be regarded as an inexhaustible mine or sink, nor is its total management, even at the planetary level, an option to be considered (see Chapter 3). We should not regard it, in Martin Heidegger's phrase, as a 'standing-reserve',[4] and should perhaps follow Paul M. Wood in distinguishing biological resources from their preconditions, such as biodiversity[5] and nature itself, which, as the source of these resources, is not a resource itself.

The ethical significance of non-humans is relevant to the question of whether resources are exclusively goods for human benefit (the conventional view). For non-human creatures too depend on their habitats and environments for food, water and the other functions, such as flight and hibernation, some even supplementing nature's supply with artefacts of their own construction such as tools, mounds or dams (and some appropriating resources intended by humans for human use). Each species has its own resources, and most themselves comprise resources for other species in complex chains of dependency. Thus resources should not be understood as essentially goods for human use. However, beyond the resources required for domesticated species, the resource requirements of non-human creatures consist mainly in intact or viable habitats; and as long as these needs are not forgotten, there is no harm in discussing resources in a context of human needs, both present and future.

In the matter of the scope and limits of resources, are nature's absorptive capacities or sinks to be regarded as resources? For some purposes these capacities have to be distinguished from material resources like food, fuel and minerals; yet these sinks resemble resources with regard both to their usefulness and to their finitude and vulnerability. Hence the capacities of the atmosphere to assimilate carbon dioxide and of the oceans to dissolve minerals are best regarded as resources. So too are the capacities of ecosystems to assimilate or recycle by-products of human activity such as pollution. Since some of these assimilative capacities are probably close to saturation, we have examples here of resources whose limits comprise a serious problem, whether or not this is the position with regard to the different kinds of material resources.

Can human beings and their skills be regarded as resources? Granted their ingenuity and their capacities for harnessing nature and thus for supplying one another's needs, this can hardly be doubted; each resourceful brain and every pair of hands can be a part of solutions, even to global problems. This is one reason why population growth is not invariably a tragedy. It is even possible to regard human labour and ingenuity, as Julian Simon does, as *The Ultimate Resource*.[6] Yet the dangers of regarding human beings as a resource are at least as great as those of regarding nature or natural systems in this light, if not greater. Human beings have moral standing, and are bearers of autonomy and other intrinsically valuable activities; indeed, for any defensible ethic, they are holders of moral rights. To regard them as mere resources is thus not only to discard any defensible form of ethics but also to deny them the possibility of the respect for which their nature and capacities qualify them. In the circumstances, without neglecting the resources comprised by human capacities for creativity, problem-solving and resourcefulness (capacities which make sustainable patterns of living possible), we do best to eschew the discourse which treats human beings as primarily resources, and to focus for present purposes on natural (material and systemic) resources instead.

SOME ISSUES ABOUT NON-RENEWABLE RESOURCES

Wherever scarce or limited resources are depleted in the present, there is an issue about provision for the future, whether the resources are non-renewable fossil fuels, such as coal, shale or oil, or renewable species or their habitats, such as forests or coral reefs, many now being not just depleted but destroyed forever. Since the future includes people all over the world, and the future of other species too, this is already not only an intergenerational issue, but also an issue with an interspecies and international bearing. Thus migratory species from habitats other than those destroyed may be put at risk; and people of Third World countries as well as of the developed world may be worse off in the future if minerals which could have benefited them (wherever they were located on the Earth) have been used up, or if their use blights global ecosystems. Recognition of landscapes, sinks and weather systems as resources will require different approaches to the related ethical issues from the approaches of theorists who have thought of resources mainly as consumables of fixed quantity to be shared within or between the generations. Non-renewable resources, however, form a good starting-point.

As Brian Barry has argued, the depletion of scarce resources carries an ethical requirement of compensation to deprived parties, which he regards as implying that similar opportunities should be available to the future as to the present generation, or rather (to be more specific) as to those societies in the current generation who consume the resources (although the compensation should not be restricted to their descendants).[7] The ethic defended in this book (see Chapter 2) upholds a similar view: current agents, to the extent that they have the necessary powers and resources, have obligations to provide for the satisfaction of the basic needs of the future, and to facilitate the development in the future of characteristic human capacities, and of the characteristic capacities of other species, to the extent that such satisfactions and development can foreseeably be facilitated. These obligations are subject to two provisos: first, the condition that basic needs of the present matter as much as like needs arising in the future, and generate comparable (and potentially conflicting) obligations (see Chapters 6, 7 and 10), and second, that future-related obligations hold only where factors beyond present control (such as climate change or future decisions) are not likely to prevent these good states of affairs from coming about. To be made effective, these obligations would need to be institutionalised in national or international agencies, authorised to tax designated resource-depleting activities, and to spend the proceeds on compensatory technology. These obligations persist even where the use of resources of the kinds currently being consumed (for example, fossil fuels) may have to be phased out after a few decades, as their future replacement will only be possible if investment takes place in the interim in research and development of new technology (such as renewable energy).

Some theorists maintain that our obligations are limited to provision for the next generation only, and do not extend to their successors. But such an approach has already been rejected in Chapter 2, and in any case hardly fits the facts of the global environment, at least where the current generation has the ability to conserve or excessively to deplete non-renewable resources and to subvert, maintain and/or restore renewable resources (sinks and ecosystems) likely to be vital for many generations to come. Even some of those sympathetic to the next-generation-only view recognise that there is a further obligation to make it possible for the next generation to conserve planetary resources and thus discharge their obligations to the succeeding generation, and thus that there is a chain of obligations for successive generations stretching into the indefinite future.[8] Without denying that people in each generation have obligations to their children, consequentialists affirm that the current generation also has obligations with regard to the needs of all their successors, insofar as the successors can foreseeably be affected in the present. These obligations stand whether or not our immediate successors shoulder their responsibilities.

However, the view that obligations with regard to the future can be grounded in consequentialism (and thus in the kind of theory advanced in Chapter 2) is sometimes confronted with the objection that our present ignorance of the future, and thus of the consequences of present actions and policies, prevents any substantive content being given to such obligations, and that a contractarian ethic of fairness (such as that of Barry), which requires generations to have equal opportunities, is thus preferable. But the problem of present ignorance of the future is actually a problem for the fairness approach (albeit not the only problem) rather than for consequentialism, for factors beyond present control, and currently unknown, could make it either prohibitively difficult or downright impossible to deliver opportunities equal to our own for the early generations of the next century, let alone for later ones; hence the inclusion in the above consequentialist principle of the proviso about factors beyond present control not ruining our efforts. The fairness approach, to the extent that it rejects consequentialism, also has to confront the problem that current agents are often held responsible (and rightly so) for the foreseeable consequences of their action and inaction, even if they adopt a theory which renounces responsibility for bad outcomes which are not actually unfair (for example, loss of species, 'compensated' possibly by funding for nature reserves).

Ignorance of the future, however, is not an insuperable problem for forms of consequentialism concerned with optimising the foreseeable outcomes of current actions, policies and institutions. Instead of requiring equal distributions of goods across the coming generations and centuries, consequentialists seek (in the light of current knowledge and uncertainty) to make the greatest foreseeable difference to the balance of value over disvalue, and to promote practices which (in the light of available information) will foreseeably have this kind of impact overall, at the same time taking into account the greater difference made by satisfying needs which are basic, and therefore prioritising this for each country and each generation into the foreseeable future. For example, a programme of investment dedicated to future energy needs, or of research into alternative sources of energy, devised to compensate for current resource-depletion would be readily justifiable on this basis, despite the uncertainties about the energy-requirements of the people at the end of the twenty-first century.[9] As we shall see, such a programme would also receive support on the same consequentialist basis in the light of the needs of the Third World in the near future as well as the further future.

But international responsibilities also arise from current resource-consumption, granted the international impact of such consumption in the rich countries of the North, the intense resource problems of the South, concerning fresh water, energy, forests, food and land, and the widely accepted obligation not to cause harm. While not all these problems can be discussed here, the general nature of this impact has to be taken into account when the ethics of resource consumption is being considered. In particular, forest, energy and fresh-water resources will be reviewed in the coming sections. Considerations of justice among members of the same generation arise as soon as issues of justice between generations do, and may also require a rather different pattern of resource access and distribution from that of the actual world; these issues will be reserved until the final section of this chapter, and returned to in Chapters 6 and 9.

WORLD RESOURCES: FORESTS

While current rates of consumption of non-renewable resources present an undeniable problem, in that they are not indefinitely sustainable, most such resources (even oil) are unlikely to be in short supply in the next few decades. Surprisingly enough, it is renewable resources and the systems on which they depend which are in the greatest danger from current consumption, resources such as forests, clean air and fresh water.[10]

Forests are much more than resources, playing key roles in climatic systems, and comprising the habitats of at least half of the species of the planet. But this very fact makes them resources too, both for their human and non-human inhabitants, and also for scientists and eco-tourists, and as reservoirs of timber, of fuelwood, and of species which humanity in general may need, not least for medical or agricultural purposes. In normal conditions forests are self-renewing, and this capacity for self-renewal is compatible with some amount of careful use such as selective logging. However, tropical forests in particular are being destroyed, at an estimated rate of 15.4 million hectares (about 0.8 per cent of their total extent) per year during the 1980s,[11] either through wholesale logging ('clear-cutting') or through burning to prepare the land for agricultural uses. Temperate and boreal forests too (for example, in Siberia) are increasingly being targeted by logging companies.

The continuing destruction of forests involves biodiversity loss on a stupendous scale, and at the same time an irreversible loss of valuable ecosystems. Although reforestation of degraded forest land has an important role for meeting future needs (and also to provide sinks for atmospheric carbon dioxide), the result is often a simpler ecosystem in which genetic resources are only partially preserved. Forestry and logging can be conducted sustainably, but often they are not. Clear-cutting, in particular, often contributes not only to deforestation but also to the erosion of soil, the degradation of watersheds, and exacerbated flooding.[12] Such outcomes are increasingly frequent, and are spreading; it is not unlikely that virtually all the lowland forests of the developing countries will have disappeared by 2020, while highland forests in places like Malaysia and southern Sumatra are also under threat.[13] Deforestation also generates widespread difficulties for the inhabitants of forests and for the people who live near them, and solutions need to take this into account.[14] But here the links between deforestation and consumption need to be addressed.

While some deforestation is ascribable to the demand in developing countries for fuel (a problem which may call for the cultivation of fuelwood as a crop), a more significant factor has been the steep increase in global consumption of wood products, consumption which doubled over the thirty years up to 1990. While the largest increase has been in developing countries, per capita consumption in industrialised countries remains greater than that of developing countries by two and a half times. This corresponds to a much increased international trade in value-added products such as paper, panels and furniture.[15] Without the underlying large and increasing demand for timber products, deforestation would be inexplicable. However, other contributory factors include the demand, largely from developed countries, for agricultural products from Central America such as beef, raised in areas cleared of forest,[16] and also the need of developing countries to find a source of revenue to service or repay their massive and accumulating debts. This factor also contributes indirectly to deforestation, when countries such as Brazil are induced by international banks to clear large tracts of forest to construct iron and aluminium mines and smelters, plus the dams, power stations, roads and railways which support them.[17] Thus consumption in industrialised countries turns out to be a major factor driving the process of deforestation, but not in isolation from the overall system of global production and finance. In other words, consumption in developed countries and some of the activities of transnational companies based in those countries are harming the forests of the Third World and the related indigenous societies; and ethical solutions must take into account the obligations of these countries and companies owed to those being harmed.

Forests are also affected by other factors: by climate change in the forms of acid rain, currently afflicting a significant proportion of the trees of Canada, Germany and Scandinavia,[18] and by global warming, which may increase tree-growth overall, but which tends to cause rapid losses of carbon when droughts give rise to forest fires,[19] as in 1997 in Papua New Guinea, Sumatra and Borneo. Acid rain could be remedied by cleaner energy generation, if producers are willing to change to cleaner processes, or if regulation is introduced by governments constraining acidic emissions (such as emissions of sulphur dioxide), or by producers turning to renewable energy sources. Global warming will be further discussed below, forests in Chapter

8, and ecological debt in Chapter 9; but one partial remedy can be presented here: the planting of new forests as carbon sinks. International agreements should encourage ecologically sensitive reforestation, whether in the North, or funded by one or more Northern countries in lands which stand to benefit, such as countries threatened by the growth of deserts.

ENERGY RESOURCES AND GLOBAL WARMING

Problems also arise from global energy consumption. These problems overlap the problems of deforestation, particularly where wood is the cheapest or the only fuel, but also extend to changes to the entire global weather system. While some partial solutions will be considered later in this chapter, the impact of energy consumption on global resources will be considered here. Projections of energy demand in the next century and proposals for overall energy solutions must in any case begin from current consumption.

Current rates of consumption of oil could produce a problem of scarcity in a few decades, as they cannot be continued indefinitely. But large-scale global problems for the planet's weather systems are almost certainly being produced in the present by global carbon emissions through the combustion of fuels and through electricity generation. Global warming is the most striking single likely effect, but there are many associated potential impacts, from the partial melting of polar ice-caps, rising sea-levels, and changes to rainfall and water-supply, to changes to winds and currents,[20] such as those involved in the El Niño effect which came to unusual prominence in 1997, and which, according to many expert commentators, was probably exacerbated by global warming, and in the more recent La Niña effect. So considerable are the changes to the global climate that projections, for example about precipitation, based on data for recent decades or on historical climate patterns are liable to be outdated. (Global warming is compatible with regional cooling, which may be happening in North Atlantic regions.) The rival explanations of global warming which seek to cast doubt on the theory that it is caused by anthropogenic emissions (for example, theories invoking sunspot activity)[21] are highly speculative, and supported by far weaker evidence than the anthropogenic theory.

Further, the rate of carbon emissions per capita of the industrialised countries (including Japan and Eastern Europe) is far higher than that of the developing countries, while the rate for North America (excluding Mexico) is considerably higher even than that of the other industrialised countries. Emission totals for China are rising, but per capita rates are lower than those for most of the Third World (except South Asia).[22] These consumption rates and the related demand for energy (principally from the industrialised countries) help to explain, among other things, the ruthless activities of companies seeking to extract oil and gas, not least in the Third World, and their willingness to co-operate with unsavoury governments that oppress or evict vulnerable minorities to open up remote areas to global industry.[23] In such cases environmental issues and human-rights issues often coalesce.

However, current effects of energy consumption and demand extend well beyond efforts to supply that demand. They also, as Hammond remarks, 'play a significant role in degrading the global commons'.[24] Some of these effects are localised, such as damage caused by oil production and refinery operations (for example, in Ogoniland in Nigeria); other effects are regional, such as acid rain, often exported from countries burning large quantities of fossil fuels (such as USA and Britain) to their neighbours (such as Canada and Scandinavia), and its impacts on trees and forests,[25] or global, such as marine pollution from oil slicks, now spread out along shipping lanes across the full extent of the oceans,[26] the growth of deserts, and the various potential outcomes of global warming already mentioned. The impacts of energy consumption on resources are largely to be found here (rather than in the direct effects of drilling), in the form of droughts and forest fires (such as those in 1997 in Indonesia, Malaysia and the Philippines), unprecedented flooding of towns, villages and farmland (as in much of East Africa then), and disruption of fisheries (as of the anchovy fishery off Peru), although overfishing is another pervasive cause of this.

Some further significant facts about energy resources should be noted, together with their implications. As Henry Shue has argued, there is currently a huge pool of unsatisfied basic human needs, and for the vast majority of humanity the satisfaction of these needs depends for the present on carbon emissions. Since global justice requires that these basic needs be met, enough carbon must be emitted and electricity generated to satisfy them.[27] However, as Greenpeace points out, if humanity uses in the coming century as much as a quarter of the known, economic reserves of fossil fuels (as opposed to estimated total fossil fuel resources), severe strains will be placed on the limits of ecological systems, and if current rates of carbon consumption continue, these limits will be severely overstretched.[28] Granted also the adverse effects of existing energy generation just mentioned, this already indicates that, unless a massive reduction of consumption can be attained by energy efficiency measures, either energy must increasingly be generated from sources other than the combustion of carbon-based fuel, or countries which already generate considerably more electricity than is required to satisfy the basic needs of their populations must generate less. I shall return to the resulting dilemmas later in this chapter and in Chapter 10.

WORLD RESOURCES: WATER

Fresh water is a finite resource, but potentially one which is renewable and which could be supplied to all human beings. Yet water pollution is currently, according to the World Bank, the most serious environmental problem facing developing countries. Well over one billion people (1.7 billion on some estimates, over a quarter of humankind) lack access to clean water, and the use of polluted water kills millions and makes over a billion people ill each year,[29] from water-borne diseases such as malaria, cholera, diarrhoea or trachoma or infestation by parasitic worms.[30] Yet urban supplies could readily be improved, granted that nearly half of all water reaching cities today is wasted through either leakage, neglect or overindulgent use.[31] Meanwhile, river and coastal ecosystems are threatened by sewage and other pollution, such as heavy metals, and runoff from the intensive use of fertilisers in agriculture.[32] Thus the problems can be traced to causes including inadequate investment in water supplies and hazardous technologies of industrial and agricultural production.

An additional global investment of $36 billion per annum (about 4 per cent of the annual world expenditure on armaments) would, it is estimated, suffice to bring to the whole of humanity clean drinking water and sanitary waste disposal.[33] Since a supply of clean water is a basic human need (and historically made a crucial difference to mortality and morbidity rates in nineteenth-century Europe), it is difficult to resist the conclusion that a global ethic would require that this sum be made available on a regular basis, whether through reduced spending on arms, or through an international tax on the arms trade, or from some other international source. In actual fact, water availability currently falls below 1000 cubic metres per person per year (a crucial benchmark for developmental and environmental problems) in a number of countries in Africa and the Middle East, and several of their neighbours are also likely to fall below this level in coming decades.[34] Indeed, by 2025 over fifty countries, with a combined population of 3 billion people, are expected to be at or around this level.[35] By the same date, Africa is expected to reach the absolute minimum of water needed per person, of 500 cubic metres, a level at which the irrigation of crops would become impossible.[36] In the Gaza strip, access is already limited to 15 gallons per person per day, and is due to fall within thirty years to 8 gallons.[37] Such intolerably low levels strengthen the ethical case for international investment in water.

In the Third World, the largest single use of water is irrigation, which has increased food production in many places, a vital prerequisite in many places of averting famine, and also a condition of self-sufficiency in others. Irrigation is the main reason for the tripling of global water use which has occurred since 1950. However, the environmental effect of hundreds of new tubewells equipped with pumps to raise underground water has often (as in parts of Tamil Nadu in southern India) been a lowering of the water-table, to the detriment of users of traditional wells and irrigation systems; lower ground-water levels (caused partly by pumping and partly by deforestation) have deprived at least 23,000 villages of water.[38] Large dams, constructed for irrigation and/or energy generation, such as the Aswan High Dam in Egypt, also have a highly questionable environmental impact, flooding traditional settlements and lands, disrupting fisheries and the deposition of silt, and silting up themselves within a few years of construction.[39]

Meanwhile, misconceived irrigation schemes of the Soviet period, combined with the diversion of rivers and intensive use of agricultural chemicals, have caused an ecological disaster in Central Asia, where the Aral Sea has shrunk to a quarter of its previous size, its dried bed now being lined with toxic substances, and where over the same period related diseases such as anaemia have become rife, and the entire regional climate has changed.[40] International efforts are now being made to restore the regional ecosystem, but may founder unless national interests can be subjected to those of the common environment.[41] Salination and water-logging are among other adverse impacts of irrigation.[42] Thus both irrigation schemes and dams require skilful management and the full involvement of local people, with their knowledge of local conditions, while large dams (as opposed to clusters of small dams) are best avoided altogether.

Climate change, mainly in the form of global warming, is exacerbating global problems of water supply and of provision for irrigation, not least when increased evaporation intensifies unsatisfied demand for water. Particular problems arise where countries such as Cambodia, Egypt, Iraq, Sudan and Syria are dependent on water originating outside their borders,[43] especially as seasonal flows are liable to fluctuate unpredictably as a result of climate change. The intensification of the global hydrological cycle resulting from global warming is likely to have many other effects on rainfall, crop fertility, rivers and lakes, effects liable to be felt most strongly in regions where water is scarce, and where there is competition for its use.[44]

A good example of these problems is the Mekong River in South-East Asia. Co-operation among the riparian countries is clearly indispensable for the communities concerned, and such sharing has been sufficiently in evidence to receive the accolade of 'the Mekong spirit'. The implicit 'community of interests' principle is an improvement on both the principle of 'absolute territorial sovereignty' (which authorises upper riparian states to treat the water in their territory as they please), and on that of 'absolute territorial integrity' (used by lower riparian states to insist that natural flows should not be tampered with). However, the 'community of interests' principle, which makes the common good of the region the determinant of distribution, is currently under threat, now that over fifty dams are being built on the Mekong. In consequence, flows and floods previously reaching Cambodia and Vietnam may dry up, and drive millions of subsistence rice farmers from the land.[45] Solutions must both be sustainable and also take fully into account the interests of the poorest, both upstream and downstream.

Sandra Postel summarises the four obligations implicit in international codes of conduct for sharing river water as follows: 'to inform and consult with water-sharing neighbours before taking actions that may affect them, to exchange hydrologic data regularly, to avoid causing substantial harm to other water users, and to allocate water from a shared river basin reasonably and equitably'.[46] These are admirable principles, and in view of the decreasing availability of fresh water per head as population increases, Postel is right to stress the need for a global ethic of sharing; but as Ted Vandeloo implies, it is crucial that talk of sharing, of refraining from harm and of equity is actually taken seriously and embodied in dealings affecting the powerless, and is not mere rhetoric.[47] It is also crucial that the richer countries mitigate global warming, and thus avoid escalation of the problems of countries whose supply is at risk (as required by the Rio Declaration).[48]

THE ETHICS OF CLIMATE CHANGE AND RESOURCE DISTRIBUTION

Climate change has complicated the issues about sharing resources between contemporaries and between generations, let alone those concerning sustainable development. Where once the way ahead for renewable resources seemed to be their management and maintenance in a renewable condition, forest ecosystems are now being subverted both by acid rain and by global warming, the absorptive capacities of the atmosphere and oceans for carbon dioxide are being overstrained, and the global layer of stratospheric ozone, with its protective capacity against skin-cancer caused by radiation, is in danger from CFCs and HCFCs.

In the case of holes in the ozone layer, no solution short of international co-operation was ever seriously possible; and as will be seen, parallel solutions are likely to prove necessary or desirable in other cases where renewable systems have been globally disrupted. Once it became clear in the 1980s that the use of CFCs was depriving humans and other species in a wide range of latitudes of their immunity from skin-cancer, agreements were reached with commendable rapidity at Montreal in 1987 that, when subsequently strengthened at London in 1990, led to an international treaty phasing out the manufacture, trading and use of these substances.[49] Even countries which had been poised to step up their use of CFCs so as to make refrigeration available to their populations were persuaded to participate, in view of offers to make available alternative technology plus funding to assist compliance, and in view of the threat to themselves as well as others posed by any such use.[50] Policing this agreement may yet prove difficult, but if that can be achieved, the problem is on the way to solution, with stratospheric ozone being expected to return to historic levels in the second half of the coming century.[51] In this way, the potentially self-renewing system of ozone protection will be restored, something likely to be applauded by people of every ethical persuasion from egoistic realists, via nationalistic communitarians to biocentric cosmopolitans. Any reservations of free-market libertarians at this global ban on a form of trade ought to be readily overcome through the reflection that a pandemic of cancer could undermine even free markets.

The control of anthropogenic carbon emissions equally requires an international solution with a view to averting major calamities and preventing the subversion of longstanding ecosystems, but the issue of which countries should bear the burden of reducing their emissions has made accomplishing this a complex task. While all countries emit carbon dioxide and other greenhouse gases, developing countries can reasonably claim that the problem arises largely from the emissions of developed countries, and that as long as energy generation depends for practical purposes on carbon sources, the Third World cannot limit its emissions, as it needs to increase them to satisfy the needs of its citizens. So developed countries are morally obliged to accept quotas and limit their emissions, without expecting quotas to be accepted in the first instance by developing countries. The Association of Small Island States has been campaigning for urgent action, before rises of sea-level inundate their territories, and the larger Third World countries, having vulnerable coastal plains, agreed to support this campaign, on condition that quotas would not initially apply to Third World countries. This position secured sufficient support among developed countries to make an agreement possible, despite strong lobbying from oil companies and the governments which to different degrees supported them, and despite attempts on the part of some developed powers to insist on targets for Third World countries.

The Kyoto agreement of December 1997 was a compromise which allocates to the various developed countries emission quotas comprising agreed proportions of their 1990 emission levels, and which also permits countries not using their quotas to trade them with countries wishing to exceed their allotted quota. It is thus based on historical emission rates; and its predicted overall effect will be to reduce the increase of emissions which would have taken place in the absence of an agreement, without halting it. The agreement includes provision for countries to meet their quotas partly by establishing carbon sinks (whether at home or overseas) which counteract above-quota emissions of theirs. Treating historical emissions as the basis of quotas may be held to reward the bigger polluters, but has so far proved the only feasible basis for an agreement. Because significant reductions from 1990 levels were accepted by the European Union, USA, Canada and Japan,[52] it can be claimed that the agreement does not simply perpetuate historical emission levels. But principles of equity continue to suggest that as negotiations proceed a more principled system should be introduced, based less on historical accidents and more on sustainable totals and on human needs.

There is no ethically defensible reason why emission quotas should be based on the state of energy consumption in 1990, apart from the usefulness of this basis in the absence of any other. Several other bases have been put forward. Steven Luper-Foy, for example, suggests that all natural resources should be regarded as available to the whole of humanity, present and future, and to be shared accordingly;[53] and while the defenders of property and of national sovereignty would regard this as highly controversial for some resources, there is a strong case for applying it to the global resource comprising the capacities of the global commons (the atmosphere and the oceans) for absorbing carbon dioxide, provided that other species do not suffer. Similarly, Michael Grubb has proposed recognition of the equal entitlement of all human beings to access to the absorptive capacities of the planet, recognition which would grant the Third World financial resources for development and environmental conservation, and simultaneously promote energy efficiency in developed countries.[54] At the same time, Henry Shue maintains that whatever global ceiling and whatever quotas are set for emissions, equity requires that provision must be made for emission entitlements which facilitate the satisfaction of everyone's basic needs, and that whatever trading of quotas is permitted, the component corresponding to emissions required for basic needs must be inalienable.[55] Consequentialists whose value-theory prioritises basic needs can endorse this stance, subject to the requirement of biocentric consequentialism that the basic needs of other species will not be subverted.[56]

Grubb's suggestion that quotas be proportional to population is one of several discussed by Finn Arler. Another possible approach would make quotas proportional to each country's absorptive capacity, but this would discriminate in favour of countries fortunate in their forests, and would be no fairer than the basis of a scaled-down proportion of historic emissions, the basis adopted at Kyoto. A further suggestion from Arler is that the costs of rationing emissions should be borne by countries in proportion to either Gross National Product (GNP) or Gross Domestic Product (GDP) per capita;[57] but such a system would have to be harnessed to some other basis for determining actual quotas, and if that basis were population, it would be strongly redistributive already, and this on more obviously relevant grounds.

The population criterion is criticised by Arler, as liable to reward and thus encourage population growth;[58] but any such tendency could be reduced by including only adults (of, say, 18 years or over). Another critic is Fidel Castro, who recognises the good intentions of the proposal, but warns that the financial resources could be largely swallowed up in debt servicing, or even in exchanges of debt-for-emission-permits, leaving Third World countries with unaltered structures and still beset by their existing problems.[59] This problem, however, could be averted if quotas for basic-needs emissions were made untradeable, as Shue advocates, and likewise emissions quotas for a quality of life slightly higher than this (geared, possibly, to secondary as well as primary education). It could alternatively be averted if most debts of Third World countries were first written off (see Chapter 9).

Even if the Kyoto agreement is ratified by the countries which negotiated it, further negotiations on emissions quotas are going to be necessary, to involve the larger Third World countries in the emerging international carbon regime, and to arrive at a sustainable carbon budget for the twenty-first century, incorporating lower carbon emissions overall than at present. Provision for reducing conventional generation through energy efficiency measures and for replacing it with renewable sources must also be prioritised, for the sake of the near future, of future generations and of other species. In such negotiations, 1990 carbon emissions cannot form the overall basis of agreement, as that basis would mean that the basic needs of millions of the world's poor would remain unsatisfied until carbon generation is replaced. Since the relevance of basic human needs is undeniable, the number of bearers of these needs will have to be introduced as a criterion. So too must inalienable carbon quotas, alongside some amount of trading in emissions permits. It will also have to be recognised, as Shue has argued, that carbon emissions which exceed one's fair share of agreed totals amount to deprivation for others,[60] and thus amount to harm that requires compensation. Accordingly, provision for monitoring compliance and enforcing compensation where appropriate is likely to be needed.

A basic-needs approach is also applicable to the issues of the availability of clean air and clean water. While changing from carbon generation and from petrol-driven transport to renewable technologies will make a large difference to air quality, separate national, regional and international efforts will continue to be needed to supply the global need for water, efforts likely to be assisted if global warming can be mitigated. This and other resource issues, however, introduce questions of national sovereignty, global justice, and intergenerational equity, which must be reserved for later treatment. They also require a discussion of sustainable development, the theme of Chapter 6.

NOTES

1. Anil Agarwal, 'Ecological Destruction and the Emerging Patterns of Poverty and People's Protests in Rural India', *Social Action and Social Trends*, January to March 1985, p. 57.
2. Alan Durning, 'Asking How Much is Enough', in Linda Starke (ed.), *State of the World 1991*, 153–69; see also Ben Jackson, *Poverty and the Planet*.
3. J. T. Houghton et al., *Climate Change 1995: The Science of Climate Change*, p. 3.
4. Martin Heidegger, 'The Question Concerning Technology', in David Farrell Krell (ed.), *Martin Heidegger: Basic Writings*, 283–322, pp. 298 and 305.
5. Paul M. Wood, 'Biodiversity as the Source of Biological Resources', *Environmental Values*, 6.3, 1997, 251–68.
6. Julian Simon, *The Ultimate Resource*. See also Principle 2 of the Program of Action of the International Conference on Population and Development, Cairo, 1994, *Population and Development Review*, 21, 1995, p. 190, which runs 'People are the most important and valuable resource of any nation.'
7. Brian Barry, 'The Ethics of Resource Depletion', in Barry, *Liberty and Justice: Essays in Political Theory 2*, 259–73.
8. Richard B. Howarth, 'Intergenerational Justice and the Chain of Obligation', *Environmental Values*, 1.2, 1992, 133–40.
9. See World Commission on Environment and Development (WCED), *Our Common Future*, 1987, ch. 7 'Energy: Choices for Environment and Development', 168–205.
10. Allen L. Hammond (ed.), *World Resources, 1994–5*, ch. 1, 'Natural Resource Consumption', p. 6.
11. Ibid., p. 7.
12. Ibid., p. 11.
13. Michael Redclift, *Development and the Environmental Crisis: Red or Green Alternatives?*, pp. 25–6.
14. Ibid. pp. 26–9
15. Hammond, 'Natural Resource Consumption', pp. 10–11.
16. Jackson, *Poverty and the Planet*, pp. 21f.
17. Ibid., pp. 23–5.
18. Hammond, 'Natural Resource Consumption', p. 17; *The Environment Digest*, no. 89/90, November/December 1994, p. 8.
19. Robert T. Watson et al., *Climate Change 1995: Impacts, Adaptations and Mitigation of Climate Change: Scientific-Technical Analyses*, p. 6.
20. Ibid., pp. 3–8.
21. Robin McKie, 'Man "Not to Blame" for Global Warming', and 'Solar Wind Blows Away Theories', *The Observer*, 12 April 1998, pp. 1 and 9.
22. Watson et al., *Climate Change 1995*, p. 42.

23. John Vidal, 'Baptism of Fire', and Ken Wiwa, 'Prime Mover', *The Guardian* (Society Section), 20 May 1998, 4–5.
24. Hammond, 'Natural Resource Consumption', p. 17.
25. Ibid., p. 8.
26. Kevin T. Pickering and Lewis A. Owen, *An Introduction to Global Environmental Issues*, 2nd edn, p. 207.
27. Henry Shue, 'Equity in an International Agreement on Climate Change' (unpublished), paper presented to IPCC workshop on 'Equity and Social Considerations Related to Climate Change', Nairobi, 1994, pp. 7–13.
28. Pete Roche, 'The Atlantic Frontier Debate: Time for Ecological Limits?', *New Ground*, Winter 1997–8, 14–16.
29. The World Bank, *World Development Report, 1992*, pp. 45–8.
30. Ted Vandeloo, 'Water, Ethics and the Global Village' (unpublished), paper presented at the conference of the Centre for Philosophy, Technology and Society on 'Ethics, Development and Global Values', University of Aberdeen, 1996, p. 3.
31. World Health Organisation, *Operation and Maintenance of Urban Water Supply and Sanitation Systems: A Guide for Managers*, pt 3.
32. Hammond, 'Natural Resource Consumption', p. 7.
33. Vandeloo, 'Water, Ethics and the Global Village', p. 4.
34. Watson et al., *Climate Change 1995*, p. 8.
35. Michael Prest, 'Water, Water, Nowhere?', *Review*, pp. 15–19; Prest here cites World Bank estimates.
36. Thomas R. Odhiambo, 'Africa Beyond Famine', in Gilbert Ogutu, Pentti Malaska and Johanna Kojola (eds), *Futures Beyond Poverty: Ways and Means Out of the Current Stalemate*, 157–64, p. 163.
37. Prest, 'Water, Water, Nowhere?', p. 17.
38. Jackson, *Poverty and the Planet*, pp. 9–10.
39. Ibid., pp. 10–11.
40. Hammond, 'Natural Resource Consumption', pp. 13f.; Prest, p. 17.
41. Max Spoor, 'Political Economy of the Aral Sea Crisis', unpublished paper presented at international Conference on 'Transformation Processes in Eastern Europe', Amsterdam, March 1997; Steve Percy, 'Arid Aral', *The New Internationalist*, 277, March 1996, p. 4.
42. WCED, *Our Common Future*, p. 134.
43. Philip Sarre and John Blunden, *An Overcrowded World? Population, Resources and the Environment*, pp. 196–201.
44. Watson et al., *Climate Change 1995*, p. 8.
45. Vandeloo, 'Water, Ethics and the Global Village', p. 6.
46. Sandra Postel, 'Forging a Sustainable Water Strategy', in Lester R. Brown et al. (eds), *State of the World, 1996*, 40–59, p. 51.
47. Vandeloo, 'Water, Ethics and the Global Village', p. 6.
48. Rio Declaration on Environment and Development, in Wesley Granberg-Michaelson, *Redeeming the Creation*, 86–90, Principles 2 and 16.

49. Houghton et al., *Climate Change 1995*, p. 3.
50. See further Chapter 10, for discussion of clashes of perspectives on this matter.
51. Houghton et al., *Climate Change 1995*, p. 3.
52. Michael Grubb, presentation on Kyoto, Royal Institute of International Affairs, London, 17 December 1997.
53. Steven Luper-Foy, 'Justice and Natural Resources', *Environmental Values*, 1.1, Spring 1992, 47–64.
54. Michael Grubb, *The Greenhouse Effect: Negotiating Targets*; also *Energy Policies and the Greenhouse Effect*.
55. Shue, 'Equity in an International Agreement', pp. 7–14.
56. See further Robin Attfield, *Value, Obligation and Meta-Ethics*. For further discussion of clashes of perspectives on this matter, see Chapter 10 below.
57. Finn Arler, 'Justice in the Air: Energy Policy, Greenhouse Effect, and the Question of Global Justice', *Human Ecology Review*, 2, 1995, 40–61.
58. Ibid., p. 56b.
59. Fidel Castro, *Tomorrow is Too Late: Development and the Environmental Crisis in the Third World*, p. 29.
60. Shue, 'Equity in an International Agreement', pp. 5–7.

CHAPTER 6

SUSTAINABLE DEVELOPMENT

In this chapter, the concepts of development and of sustainability will be introduced, with a view to considering whether the theory of sustainable development, which has been put forward as supplying solutions in outline to ecological problems and also to the problem of poverty, represents a coherent position, and whether it stands up to ethical and practical objections. I shall argue that both the conceptual and the substantive problems can probably be overcome, but that the attainment of sustainable development, however indispensable, may involve restructuring the world.

THE CONCEPT OF DEVELOPMENT

The term 'development' has many senses, clustering around the sense of the realisation of capacities whether in protozoa, plants or individual people. But the advocates and the critics of development of recent decades understand it alike as a state of society or a social process with a moderately specific content, albeit one which leaves room for disagreement about the criteria of development and the best path to take in its direction. In the light of recent usage, development is here understood to be a social state or process which is present when the interconnected evils of underdevelopment (poverty, disease, illiteracy, high infant mortality, low life expectancy, low productivity and poor medical and educational facilities) are reduced or averted and replaced by the attainment of health, literacy, low infant mortality, higher life expectancy, higher productivity, good medical and educational facilities and sufficient prosperity to allow the evils of underdevelopment to be held at bay. The process of development is not to be identified with economic growth; the defining characteristic of a developing society would rather be the satisfaction of basic needs, including needs for individual automony and for making meaningful contributions to society. Hence the process of development will also tend to eliminate the more blatant inequalities, and typically involves self-determination and self-help on the part of the society as a whole, and active participation on the part of most of the individual members.[1] The Brundtland Report, which represents the meeting of basic needs as the core of development,[2] clearly presupposes a concept of development similar to the one just delineated.

Not only are economic growth and development distinct; they can even conflict, particularly where sustainable development is in question (see below). Thus economic growth can fail to satisfy basic needs, and can increase inequalities, while meeting these needs can involve constraining growth, whether by regulation, social planning or public accountability. One model of development, the trickle-down theory, maintains that development is nevertheless promoted by economic growth and its tendency to spread across societies and the world economy; manifestly the facts of the contemporary world do not fit this theory. Indeed it is far from clear that development need involve the consumerism or the market values which are so pervasive in the societies of the 'developed' North. However, it is because of theories which either identify growth and development, or yoke them as indispensable cause and characteristic effect, or represent development as movement towards consumerist economies, that many of the criticisms of development shortly to be encountered arise.

It should also be noted that 'development' is not actually defined here as desirable or as improvement, or made into a prescriptive term. I can endorse the United Nations account of development, in the *Declaration on the Right to Development* (1986), as:

> a comprehensive economic, social, cultural and political process, which aims at the constant improvement of the well-being of the entire population and of all its inhabitants on the basis of their active, free and meaningful participation in development and in the fair distribution of the benefits resulting therefrom,[3]

subject to the inclusion of the aim of the reduction of poverty; but this account conveys that development aims at improvement and so on, without making all development essentially desirable, for aims are not always on target. Further, if the concept of social justice is partly to be analysed in terms of the satisfaction of basic needs (as I have contended elsewhere),[4] then a process which brings about the satisfaction of unsatisfied needs has to be recognised as a move towards justice (unless the needs of other species are disregarded), and thus as desirable; but this recognition depends on an understanding of justice, rather than of development. For the account given here, the concept of development has a factual, objective core, which does not automatically settle the issue of the desirability of development, and which also leaves plenty of room for disagreement about the best way forward for given societies. By the same token, criticisms are inapplicable to this account of development which say that development theory makes development invariably a favourable change, whatever is developed being necessarily superior (including the societies of the North), and whatever is underdeveloped inferior (including the Third World).[5] Again, this account allows objections to development as undesirable to be voiced, rather than making them incoherent.

This account also provides for diverse paths to development, albeit with a common core. The satisfaction of human needs allows of a wide range of cultural expressions, varying both between and within societies, and thus of the flourishing of minority cultures, and is consistent with diverse values, such as those of different religions. No uniform pattern of development is mandated by this theory of development, something which it shares with the somewhat differently conceptualised account of Dower.[6] It is thus compatible with the advocacy of endogenous development,[7] or development which takes into account the particularities of each nation, and is driven largely by local values, as long as the satisfaction of basic needs continues to be prioritised. Indeed both cultural diversity and minority rights are championed by the kind of consequentialism defended in this book. The contradiction feared by Gustavo Esteva, namely that the whole notion of development requires a uniformity of approach inconsistent with any processes driven by endogenous values, thus proves unreal, even though cases of imposed uniformity in the name of development sometimes occur. As long as the high priority of the satisfaction of basic needs, including the need for individual autonomy, is recognised, the consequentialist ethic which I defend supports a wide range of developmental processes, each best suited to different historical and cultural situations.

Esteva supplies a history of the concept of development, in vindication of his plea to reject it. He finds it to arise from the beliefs of philosophers such as Herder in the automatic and homogeneous unfolding of history, and of Marx in the necessary character of history's laws, and then to have been taken over by Harry Truman in 1949 in the name of a programme of development to bring progress to the 'underdeveloped areas' of the globe, but in fact to underpin American hegemony.[8] We can set aside here questions of the interpretation of Herder, Marx and Truman; the point is rather that the concept of development in the present need have no overtones either of inevitability or of homogeneity (even if it sometimes had them in the past), particularly when it is expounded in terms of self-determination and autonomy, and equally need not carry overtones of Truman's beliefs about material progress. We can grant to Esteva that where particular patterns of development are imposed by Western governments, global financial institutions, or global business, there are grounds to question them, but that is no reason for rejecting the entire concept of development. While characterisations of two-thirds of the world as 'underdeveloped' are clearly out of place, the notion of underdevelopment is in place to depict the avoidable loss of human potential implicit in the often interconnected evils of poverty, illiteracy, malnutrition, disease and premature death. And while so-called 'developed' societies have much to learn from Third World societies, development remains important wherever characteristic human capabilities remain unrealised.[9]

THE CONCEPT OF SUSTAINABILITY

Before sustainable development is discussed, it is worth remarking that proposals close to sustainability had been independently advocated by environmentalists, systems theorists and economists long before the Brundtland Report famously paired sustainability and development, and urged their joint pursuit.[10] For example, in 1977 Herman Daly and others had advocated *The Sustainable Society*, a society with stabilised levels of population and production,[11] while in 1972 in *Limits to Growth* Donella Meadows and others had urged upper limits to population, capital and pollution.[12] Mary B. Williams had even put forward a theory of how to manage renewable biological resources sustainably, through co-ordinated self-restraint.[13] The sustainability of a process, practice or society is its capacity to be practised or continued indefinitely; and these writers were conveying that society could be perpetuated only if ceilings were to be observed for certain kinds of growth.

So the juxtaposition of sustainability and development on the part of the World Commission on Environment and Development (WCED) (and of the *World Conservation Strategy* which preceded it in 1980)[14] seemed an oxymoron if not a contradiction. For those who assumed that development and growth could be equated, the appearance of contradiction was strong, as mention of sustainability conveyed limits, and evoked the impossibility of (at least) perpetual growth. Even if 'sustainability' was intended as an epithet qualifying 'development' and did not express a separate goal for society, the values conveyed included a new stress on intergenerational equity and long-term present responsibilities towards future generations, and also, despite its omission from the opening definition of 'sustainable development', recognition by the Commission of the non-economic value of the environment, of human dependence on its continuing intactness, and in a few passages of the moral standing and intrinsic value of non-human living creatures.[15] Sustainable development is thus development which is sustainable both economically and environmentally, and where these considerations are integrated in actual policies.

However, global sustainable development was in actual fact held by WCED to be consistent with economic growth in places where human needs are not being met, particularly in the Third World,[16] growth there being held to be necessary for development. The limits recognised were 'not absolute limits' but 'limitations imposed by the present state of technology and social organization on environmental resources' and 'by the ability of the biosphere to absorb the effects of human activities'.[17] Thus sustainable development is consistent with more responsible and more efficient use of resources which could still facilitate selective growth; at the same time global sustainable development involves the more affluent societies curtailing their energy use so as to 'adopt life-styles within the planet's ecological means'.[18] Indeed, in the absence of global sustainability, national or regional sustainability could easily be undermined; hence sustainable development has to be sustainable not only economically and environmentally but globally too.

The World Commission's opening definition of sustainable development runs as follows: 'development that meets the needs of the present without compromising the ability of future generations to meet their own needs'.[19] One major implicit theme is that well-being should not decline over time, and that future generations should be enabled to meet their own needs through being compensated for resource consumption in the present, thus receiving equivalent resources (environmental resources included) to those received by the current generation. For one rather optimistic interpretation, this suggests that if each generation compensates its successor in this way, all future generations will be provided for, and that it is largely unnecessary for present people to do more than this for generations after that of their successors; recognised exceptions include cases where present action has long-term effects such as where biodiversity is irreversibly lost for all generations.[20] But in practice WCED seeks on several fronts to plan for more generations than one, for example with regard to the energy needs of the middle decades of the next century;[21] such an approach accords better with the underlying concept of sustainability, and also receives the general support of a consequentialist ethic (and its particular support when a low-energy future based on energy efficiency, use of renewable energy sources, and increased Third World consumption is advocated). Principles of intergenerational equity are discussed further in Chapter 10 (below).

Another central theme of WCED concerns the need to take fully into account the costs of environmental damage and destruction; only in this way can economic processes be geared to sustainability.[22] In part, this reflects the widespread concern of environmental economists that costs previously regarded as external to balance-sheets (externalities) should be internalised within accounting systems, and that decision-makers should thus be required either to justify them or to seek to avoid them. In part this theme extended to the proposal that institutions such as the government ministries accused of causing environmental deterioration should be given the additional brief of preventing it. In this way, instead of needing to repair such damage, governments would seek to anticipate and pre-empt it. WCED applies this to international agencies too, such as the International Monetary Fund (IMF) and the World Bank, with their questionable record in environmental matters.[23]

Problems, however, are raised by the notion of inheriting resources equivalent to, or substituting for, those inherited by a previous generation (as all this involves valuations of nature, and also some resources may be irreplaceable), and these become thornier when what is inherited is interpreted by expositors of sustainable development as 'environmental wealth'[24] or as 'natural capital',[25] for wealth and capital are property, and elements in the worldwide property system, which would thus expand beyond the realm of culture to the furthest reaches of nature (see below). These problems carry particular importance in the light of the international endorsement accorded to sustainable development at the Rio Summit (1992),[26] and need to be considered in turn.

Given its commitment to meeting current needs and using resources to do so, some of them non-renewable, WCED is committed to there being goods valuable enough to justify some consumption of resources; its commitment to compensation of succeeding generations further commits it to there being compensatory goods comparable in value to those consumed. The forms of metric for these values employed by environmental economists (such as the sums of money which respondents to questionnaires say that they would be willing to pay to preserve various natural items) are open to all kinds of difficulties (to say the least), and these difficulties lend credibility to those who suggest that the different values involved in environmental gains and losses, compensation and substitution, are incommensurable and cannot rationally be compared.[27] Yet if we hold (as Brundtland clearly does) that it is sometimes justified to build a cluster of houses to meet people's need for accommodation even in a rainforest, where any tree-felling is liable to destroy several localised species as well as particular trees, we thereby hold that there is sometimes enough reason or value in favour of doing this to outweigh the value that is lost. And even if we hold that on balance this would never be justified, we are still committed to comparing and balancing reasons and values. So it is not incoherent to compare the values concerned, even if no satisfactory metric or system of environmental weighing exists.[28]

However, there are clearly some limits to the possibility of compensation for environmental depletion or damage. No amount of new technology, however benign, could compensate humanity or other species with similar vulnerability for the loss of the ozone layer, let alone for the loss of an ocean or continent, rendered uninhabitable, perhaps, through ozone depletion. And although there are problems in declaring ecosystems irreplaceable (partly because scientists encounter difficulties in identifying any such stable systems, and locating their boundaries and extent),[29] sometimes the combined intrinsic value of their members and their instrumental value for communities which depend on them would justify a prohibition on eroding or extinguishing them, as consequentialists can readily acknowledge.[30] Adherents of sustainable development diverge over whether there are limits to the possibility of substitution;[31] but if there were no limits, then sustainable development would cease to embody environmental sustainability. Thus if what is to be sustained and if possible perpetuated is not just the more material aspects of human well-being, but also the ecosystems on which most species depend and the well-being of the full range of morally considerable creatures (or even most of that range), then sustainability requires firm upper limits to substitution, and so does sustainable development. Thus sustainable development has been redefined in a United Nations Environment Programme report as: 'Improving the quality of life while living within the carrying capacity of supporting ecosystems',[32] a definition which, without being all-sufficient (particularly where ecosystem support for human populations is concerned), usefully supplements that of Brundtland.

Sustainable development also throws into question the characteristic practice of economists of discounting all future costs and benefits exponentially by a fixed annual rate (the social discount rate), although some environmental economists endorse it.[33] The central justification of this practice is sheer time-preference, or the preference of consumers for present over future benefits; but acting on this basis is bound to disadvantage the future (and effectively to disfranchise altogether the future of more than a few decades from the present). Other justifications are also put forward, such as opportunity costs (the costs of foregoing present opportunities) and the uncertainty of future impacts; recent philosophical appraisals of such grounds, however, conclude that they fail to justify across-the-board discounting, as opposed to selective discounting in the particular cases where such grounds apply. The same applies to justifications such as the undesirability of excessive burdens falling on any one generation (such as the present generation); while the urgency of present needs and the limits of the possible justify ceilings to present self-sacrifice and a sharing of burdens between generations, no justification is to be found for across-the-board discounting here either.[34]

Nor is such a justification to be found in our special relations to specific contemporaries. While one kind of communitarian would make such relations ethically central, and responsibilities to those more distant in space or time peripheral, this position is clearly unacceptable to environmentalists concerned for the global environment and for generations stretching into the future. Other communitarians, such as Avner de-Shalit, endorse ampler responsibilities to the future, grounded on the needs and values of those now alive.[35] Consequentialists, however, as was argued in Chapter 2, can recognise ties of family, community and country (beneficial institutions all, within limits) without the related obligations annulling or invariably overriding responsibilities to avoid harm and foster well-being among all bearers of moral standing, whether or not they belong or will belong to our community.[36]

Another proposal of economists supportive of sustainable development is that natural resources be recognised as environmental wealth or natural capital; only in this way, they hold, will it be treated with sufficient care, rather than being treated as a bottomless mine or inexhaustible sink.[37] People will then be motivated to preserve the natural environment as they might the family silver. But, as Martin O'Connor has pointed out, this is part of an attempted expansion of the property system so as to appropriate for it not only those outlying societies which were hitherto outside it but also the entire natural order,[38] which (as has been seen) cannot be owned. To suggest that it can be owned, I want to add, is anthropocentric to the point of absurdity. Presumably on the day that humanity becomes extinct, nature either suddenly becomes ownerless, or reverts to being common ground for all the species, as, until the theory of natural capital was put forward, it was customarily assumed to be already. In any case this kind of commodification of nature should be resisted, just like the commodification of contemporary pre-market societies. In some forms, which designate certain natural items as critical natural capital that, as such, are beyond price and outside the global auction, it may in practice be harmless; but that does not make it conceptually defensible. Fortunately, however, adherents of sustainable development have no need to commit themselves to belief in natural capital; the only commitments needed by the theory in this regard are belief in responsible treatment of resources (absorptive capacities included) and in limits to their substitutability. Indeed, belief in the independent value of living creatures, as has been seen, is entirely consistent with their stance.

DOES SUSTAINABLE DEVELOPMENT HARBOUR CONTRADICTIONS?

The Rio summit Declaration adopted sustainable development, albeit in an entirely anthropocentric interpretation,[39] and the signatories to Agenda 21 each agreed to prepare national strategies for sustainable development by 2005.[40] So it is unsurprising that most institutions nowadays profess allegiance to sustainable development, while often using it as a cloak for business as usual. Politicians, for example, often use the phrase to convey nothing more than economic growth for as little as a decade. Accordingly the rhetoric of sustainable development (like that of democracy) harbours whole flotillas of contradictions, wherein what speakers favour frequently clashes with either the international or the intergenerational or the environmental aspects of development which is genuinely sustainable.

This would not greatly matter if the central concept of sustainable development (and its many implicit goals) is capable of consistent application; but serious doubts have been expressed in this regard. Thus Henry Shue draws attention to the potential conflict between trade liberalisation (which Duncan Brack shows to be thought to promote sustainable development)[41] and environmental protection, for example when fisheries are permitted to disregard their capture of dolphins.[42] In this connection, Shue stresses that pairs of institutions professing different desirable aims cannot be guaranteed to deliver these goals in combination.[43] (One reason for the potential conflict could be that trade liberalisation tends to strengthen the very companies and practices liable to undermine both local economies and local ecosystems.) Writing before Kyoto, Shue also questions the claim of the OPEC nations that 'poor nations can develop, rich nations can continue to expand their economies, and climate change can be avoided without any initiative to move away from fossil fuels as our main energy source', adding the comments that 'if "development" means industrialisation based on fossil fuel, there is no such thing as "sustainable development"', and that choices between values will have to be made.[44] Fortunately Kyoto marked the beginning of a realisation that growth among rich nations cannot depend on a growing use of fossil fuels; but a full recognition that sustainable development cannot indefinitely be based on fossil fuel is yet to be achieved.

Paul Ekins expresses parallel doubts. Without rejecting belief in sustainable development, he specifies demanding conditions for its realisation, and presents an argument for doubting the possibility of combining it with growth. Ekins borrows from Paul and Anne Ehrlich the formula which makes environmental impact (I) the product of population (P), consumption per capita, or affluence (A) and intensity of consumption, or technology (T).[45] The Ehrlichs' IPAT formula is also cited with approval by Nigel Dower.[46] It requires some qualification, as environmental impacts are not always incremental, can suddenly become disastrous when thresholds are crossed, and are often quite independent of size or growth of population (see Chapter 7), but is helpful, for example, with regard to energy consumption.

Ekins' point is that if, during the coming fifty years, the human population doubles, as some projections predict, and if consumption quadruples, as a moderate growth rate would indicate, and if impacts need to be halved to attain sustainability (a more moderate reduction than IPCC suggests), then T (intensity of use per unit of resources) would need to increase to a multiple of sixteen of current levels. Since this is very unlikely to be attained, and its non-attainment would mean that sustainable levels of impact may not be achievable together with growth, then sustainable development and growth cannot readily be combined.[47] The moderate rate of growth mentioned by Ekins (2 to 3 per cent) is in fact lower than that advocated by WCED for the Third World, and may be no less than is needed to satisfy basic needs. Thus Shue's point about the need to choose between the values associated with sustainability and development seems to be borne out.

However, population may well stabilise at less than double the current population (see Chapter 7). Also efficiency in resource-usage has been growing by leaps and bounds.[48] So it may be possible to attain some amount of growth in the Third World (growth necessary to facilitate development there, as Ekins recognises)[49] without unsustainable overall impacts. What Ekins' argument shows, however, is that it is exceedingly difficult to combine this with economic growth in rich countries, with their high per capita consumption, even if their populations remain static. Rather, as he concludes, Northern countries and institutions must take radical steps to rectify their global environmental impact (or 'footprint'), and also to reform the current structures of aid, trade and debt which make sustainable development in the South impossible.[50]

Ekins' argument does not show, and is not intended to show, that sustainable development is impossible; and, unlike many of the contributors to the same volume, he does not regard development itself as undesirable. But his argument does show that sustainable development is virtually impossible with the world structured as it is. If so, it is not surprising that those who assume the continuation of these structures fall into contradictions when they advocate sustainable development at the same time.

This assumption also accounts for the contradictory specifications of sustainable development noted by John S. Dryzek, whose view is that sustainable development is 'nowhere an accomplished fact, save in small-scale hunter-gatherer and agrarian societies' but is rather a globally dominant discourse, open to appropriation from antithetical political camps.[51] Through acknowledging that it can assume material form, Dryzek recognises the consistency of the notion, despite the problematic nature of his examples; and while he is right about much sustainability discourse, appeals to sustainable development are immensely more credible when they take the above definitions seriously. Alerted by writers like Dryzek, we should be ready to disown many (possibly most) of the proposed specifications of sustainable development; this helps to clear the way for sustainable development itself, and the necessary related project of global restructuring along lines advocated by Ekins.

OBJECTIONS TO SUSTAINABLE DEVELOPMENT

Sustainable development has been a target of ethical and related criticism from at least two directions. Where it sets firm upper limits to substitution, comprising (in Wilfred Beckerman's terms) 'strong sustainability', it is criticised by Beckerman as morally repugnant, for countenancing the preservation of all species, including beetles, even at the cost of a failure to deploy resources to alleviate widespread poverty and environmental degradation.[52] (Beckerman's counterpart criticism, that the alternative interpretation of sustainable development, which permits unlimited substitutability, involves no significant departure from conventional economics,[53] requires no further discussion here, but serves to underline his challenge to sustainable development to justify its distinctive elements.) But the very different (though equally radical) critique of Wolfgang Sachs and fellow essayists, that sustainable development theory embodies the co-option of environmentalist protest by Western establishments, intent on imposing a uniform global hegemony,[54] also merits consideration. While the concept of sustainable development does not incorporate all authentic values and is not immune from ethical criticism,[55] its applicability to global problems depends on adequate defences against criticisms such as these.

On Beckerman's criticism regarding the preservation of beetles, Michael Jacobs justifiably replies that strong sustainable development need not require that all species be preserved. Committed as its adherents are to environmentally sustainable development, and thus both to environmental improvements for the poor and for the preservation of most ecosystems and species, they have to be prepared for conflicts between these commitments, and the preservation of a given endangered species will not always then receive the highest priority, particularly if the conflict is with unsatisfied needs of poor people for (say) clean water. Further, sustainable development does not comprise a comprehensive and exhaustive ethic, and its adherents may be committed also, without inconsistency, to broader principles of justice requiring the needs of the poor or of future generations to be prioritised. Also there will often be no conflict between preserving species and enhancing the situation of the rural poor, for the kinds of solutions which give local people a stake in preservation often exemplify ways of achieving both.[56]

While these responses can be endorsed, they need to be supplemented. While sustainable development cannot require preservation to take invariable priority over present human needs,[57] it will sometimes involve the preservation of a species or a habitat or a climatic system receiving priority, whether for the sake of future human generations, or to prevent climate change, or for the sake of vulnerable creatures and species. In such cases it is likely to diverge from policies regarded as optimal by conventional cost/benefit analysis, either because it does not discount future benefits so frequently or so heavily, or because it takes into account a broader range of human interests, or because it takes the interests of non-humans seriously.[58] Or its adherents may be concerned to refashion policies by which an apparently sustainable society exports unsustainability in the longer term to others. But these considerations, far from being morally repugnant, are important ethical correctives to the pursuit of optimising the balance of benefits and costs as conventionally understood.

Before the relations of sustainable development to the wielders of international power are considered, the criticisms of Donald Worster (a fellow essayist of Sachs) should be addressed. Worster's claim that sustainable development treats nature as 'nothing more than a pool of "resources" to be exploited' and as having 'no intrinsic meaning or value'[59] fits none but anthropocentric interpretations of sustainable development, but provides a salutary warning against the attenuation of the concept in practical situations, of which biocentric readers should remain vigilant.

However, Worster's criticisms that sustainable development theorists too readily assume that we can easily identify ecosystems and determine their carrying capacity, granted recent tendencies in ecological science to be less confident than previously about the supposed order and determinacy of natural systems,[60] generate more serious problems for sustainability theory. The usual assumption about the care of renewable resources has been that we can determine the maximum sustainable yield from, for example, a given fishery, and that by fishing within these limits can both preserve fish populations and harvest this yield indefinitely (as Mary Williams has argued).[61] If, however, fishstocks and other renewable resources fluctuate too greatly, this policy (even if strictly observed) will instead generate disastrous overfishing, an effect, Worster believes, it actually is having. Relatedly, theorists who claim to identify the 'critical loads' that given systems will bear in the cause of sustainable development, but on the basis of insufficient understanding of the systems, may actually contribute to the collapse of what they purport to sustain.[62]

Theorists of sustainability should give ground to these criticisms, without abandoning their overall position. In view of the limits to our knowledge of resources and of systems, greater margins should be left to allow for error, and greater effort should be made to understand the systems comprising our environment, and their limits. Greater effort should also be made to consult local people whose understanding of systems is too easily overlooked. Planning which heeds such ampler margins for error and enhanced understanding may then be able to forestall further instances of overfishing and comparable disasters. Thus ecological limits need first to be studied and as far as possible established before decisions about resource management are made,[63] as the UNEP definition of sustainable development requires. Granted that the flaws specified by Worster can be corrected, and are not intrinsic to sustainable development, the underlying goals of sustainable development make it too important to be abandoned; nor should preservation be given a general priority over development, as Worster seems to suggest.[64] Such a policy would be vulnerable to the charge of culpably neglecting poverty implicit in Beckerman's critique.

When Worster moves to criticising sustainability for an uncritical acceptance of secular materialism and belief in progress, he comes closer to Sachs' charge that environmentalists supportive of sustainable development have become instruments of the global establishment. The narrower charge of uncritical secular materialism turns out to be wide of the mark, if (as by now seems likely) sustainable development can be harnessed to biocentric cosmopolitanism and to a metaphysic of stewardship, which (as was argued in Chapter 3) can be held in either a religious or a secular form. Indeed the charge of uncritical pursuit of progress misfires against a stance which rejects unsustainable policies of resource usage, production and consumption even more resoundingly than Esteva's corresponding charge against the concept of development. Yet the very endorsement of sustainable development by Western governments, as at the Rio Conference, and also by international banks, raises understandable fears that it has been appropriated by the forces of economic orthodoxy. The disappointment with the outcomes of the Rio conference in matters of climate change and of biodiversity supplies apparent evidence for this interpretation,[65] as does the yet stronger disappointment with the Rio-plus-Five conference, held in New York in 1997.

But the adoption of some of the central themes of the Brundtland Report by the international conference which it called for cannot be allowed to discredit those themes, even if leading politicians and bankers endorsed them more by way of rhetoric than out of conviction, and even if this very adoption destined them to dilution. Crucial concepts and telling theories do not become incoherent or void through partial acceptance by the powerful; and sometimes their acceptance betokens their near indispensability (as in the matter of global warming). If through this acceptance they come to be used to impose a centralised, homogenising model, whether of development or of sustainability, this trend is open to legitimate criticism not only from critics of globalisation but also from the advocates of sustainable development itself, who can now appeal both to works like *Our Common Future* and to the Rio Declaration itself, with its assertion of entitlements to autonomous sustainable development and to national sovereignty.[66] The failure of the international system to undergo whatever radical reform or restructuring these themes require does not begin to show that they were all along misguided, whether ethically or politically.

These defences of the theory of sustainable development do not show, and are not intended to show, that it embodies a comprehensive solution to the problems facing humankind. For example, *Our Common Future* may be too optimistic about the prospects for combining sustainability and growth, at least in the North, as Ekins suggests. Further problems (some considered in later chapters here) will disclose a need to supplement the solutions there proposed, as will further reflection on values such as justice. There again, sustainable development may be genuinely the way forward, but only if the structures of international aid, trade and debt are transformed out of recognition (as Ekins also suggests), and only if the countries of the South co-operate to preserve their biodiversity and to research pathways for biotechnical development (as Castro proposes),[67] and perhaps only if international capitalism is either superseded or subjected to global regulation to curtail its inherent tendencies to unsustainability (as James O'Connor argues).[68] If so, then solutions may still need to incorporate sustainable development, but are largely to be found elsewhere.

NOTES

1. A similar but fuller account of development is offered in Robin Attfield, 'Development: Some Areas of Consensus', *Journal of Social Philosophy*, XVII, Summer 1986, 36–44. On basic needs, see n. 4 (below).
2. World Commission on Environment and Development (WCED), *Our Common Future*, pp. 8 and 43–4.

3. United Nations, *Declaration on the Right to Development*, preamble, para. 2.
4. Robin Attfield, *Value, Obligation and Meta-Ethics*, ch. 9. The notion of basic needs is expounded there in chs 5 and 6.
5. See Gustavo Esteva, 'Development', in Wolfgang Sachs (ed.), *The Development Dictionary*, 6–25, p. 10.
6. Nigel Dower, *World Ethics: The New Agenda*, chs 6 and 8.
7. Esteva, 'Development', pp. 15–16.
8. Ibid., pp. 6–10.
9. See the various contributions to Martha Nussbaum and Jonathan Glover (eds), *Women, Culture and Development: A Study of Human Capabilities*.
10. WCED, *Our Common Future*.
11. Dennis Clark Pirages (ed.), *The Sustainable Society: Implications for Limited Growth*.
12. Donella Meadows et al., *The Limits to Growth*.
13. Mary B. Williams, 'Discounting versus Maximum Sustainable Yield', in R. I. Sikora and Brian Barry (eds), *Obligations to Future Generations*.
14. International Union for the Conservation of Nature, *World Conservation Strategy*.
15. WCED, *Our Common Future*, pp. 57, 147, 155, 163.
16. Ibid., p. 44.
17. Ibid., p. 8.
18. Ibid., p. 9.
19. Ibid., p. 43. There are now some 300 further definitions either of sustainability or of sustainable development: see Andrew Dobson, 'Environmental Sustainabilities: an Analysis and a Typology', *Environmental Politics*, 5.3, 1996, 401–28, p. 402.
20. David Pearce et al., *Blueprint for a Green Economy*, pp. 3, 34–6.
21. WCED, *Our Common Future*, ch. 7: 'Energy: Choices for Environment and Development'.
22. Ibid., p. 37.
23. Ibid., pp. 39, 337–8.
24. Pearce et al., *Blueprint for a Green Economy*, p. 3.
25. Ibid., p. 3
26. United Nations Conference on Environment and Development, 'Rio Declaration on Environment and Development', in Wesley Granberg-Michaelson (ed.), *Redeeming the Creation*, 86–90.
27. John O'Neill, *Ecology, Policy and Politics: Human Well-Being and the Natural World*, 102–15
28. It is argued in Robin Attfield and Katharine Dell (eds), *Values, Conflict and the Environment*, that a form of environmental weighing is possible, in which all the relevant values, including future interests and non-human interests, are taken into account. My qualifications vis-à-vis the central argument are explained in Robin Attfield, *Environmental Phi-*

losophy: Principles and Prospects, ch. 11 ('Reasoning about the Environment').

29. See Donald Worster, 'The Shaky Ground of Sustainability', in Wolfgang Sachs (ed.), *Global Ecology: A New Arena of Political Conflict*, 132–45.
30. Attfield and Dell, *Values, Conflict and the Environment*, pp. 65–6.
31. See Wilfred Beckerman, 'Sustainable Development: Is it a Useful Concept?', *Environmental Values*, 3.3, 1994, 191–204, pp. 194–5; also Herman Daly, 'On Wilfred Beckerman's Critique of Sustainable Development', *Environmental Values*, 4.1, 1995, 49–55.
32. United Nations Environment Programme et al., *Caring for the Earth*.
33. Pearce et al., *Blueprint for a Green Economy*, ch. 6; William Nordhaus, *Managing the Global Commons*, p. 10.
34. Derek Parfit, 'Energy Policy and the Further Future: The Social Discount Rate', in Douglas MacLean and Peter G. Brown (eds), *Energy and the Future*, 166–79; also *Reasons and Persons*, pp. 480–6; John Broome, *Counting the Cost of Global Warming*.
35. Avner de-Shalit, *Why Posterity Matters: Environmental Policies and Future Generations*.
36. See further Robin Attfield, 'Discounting, Jamieson's Trilemma and Representing the Future', in Tim Hayward and John O'Neill (eds), *Justice, Property and the Environment: Social and Legal Perspectives*, 85–96.
37. Pearce et al., *Blueprint for a Green Economy*, pp. 41f.
38. Martin O'Connor, 'On the Misadventures of Capitalist Nature', *Capitalism, Nature, Socialism*, 4.3, September 1993, 7–40.
39. Rio Declaration, Principle 1; Granberg-Michaelson, *Redeeming the Creation*, p. 86.
40. See Cmd. 3789, *Eliminating World Poverty: A Challenge for the 21st Century* (White Paper on International Development), 3.39, p. 65.
41. Duncan Brack, 'Balancing Trade and the Environment', *International Affairs*, 71.3, 1995, p. 497.
42. Ibid., pp. 501–2.
43. Henry Shue, 'Ethics, the Environment and the Changing International Order', *International Affairs*, 71.3, 1995, 453–61, p. 460.
44. Ibid., pp. 460–1.
45. Paul Ekins, 'Making Development Sustainable', in Wolfgang Sachs (ed.), *Global Ecology*, 91–103, pp. 92–3.
46. Dower, *World Ethics*, ch. 9.
47. Ekins, 'Making Development Sustainable', pp. 92–3.
48. See Mark Sagoff, 'Carrying Capacity and Ecological Economics', in David A. Crocker and Toby Linden (eds), *Ethics of Consumption: The Good Life, Justice and Global Stewardship*, 28–52.
49. Ekins, 'Making Development Sustainable', p. 95.
50. Ibid., p. 99.

51. John S. Dryzek, *The Politics of the Earth: Environmental Discourses*, pp. 123–8.
52. Beckerman, 'Sustainable Development', pp. 194f.
53. Ibid., pp. 195, 199–201.
54. Wolfgang Sachs, 'Global Ecology and the Shadow of "Development" ', in Sachs (ed.), *Global Ecology*, 1995, 3–21; see also the essays in this collection by Achterhuis, Shiva and Worster.
55. See Robin Attfield and Barry Wilkins, 'Sustainability', *Environmental Values*, 3.2, 1994, 155–8.
56. Michael Jacobs, 'Sustainable Development, Capital Substitution and Economic Humility: A Reply to Beckerman', *Environmental Values*, 4.1, 1995, 57–68, pp. 62–3.
57. Robin Attfield, 'Saving Nature, Feeding People and Ethics', *Environmental Values*, 7, 1998, 255–68.
58. This point is made in the language of limits to substitution in Richard Dubourg and David Pearce, 'Paradigms for Environmental Choice: Sustainability *versus* Optimality', in Sylvie Faucheux, David Pearce and John Proops (eds), *Models of Sustainable Development*, 21–36.
59. Worster, 'The Shaky Ground of Sustainability', p. 142.
60. Ibid., pp. 141–2.
61. Williams, 'Discounting versus Maximum Sustainable Yield'.
62. Worster, 'The Shaky Ground of Sustainability', p. 142.
63. Bryan Norton, *Toward Unity among Environmentalists*, p. 118.
64. Worster, 'The Shaky Ground of Sustainability', p. 143.
65. J. Martinez-Alier, 'Distributional Obstacles to International Environmental Policy: The Failures at Rio and Prospects after Rio', *Environmental Values*, 2.2, Summer 1993, 97–124, p. 98.
66. Rio Declaration, Principles 1–6; Granberg-Michaelson, *Redeeming the Creation*, pp. 85–6.
67. Castro, *Tomorrow is Too Late*, p. 51.
68. James O'Connor, 'Is Sustainable Capitalism Possible?', in Martin O'Connor (ed.), *Is Capitalism Sustainable? Political Economy and the Politics of Ecology*, 152–75; see also Tim Hayward, *Ecological Thought: An Introduction*, p. 119.

CHAPTER 7

POPULATION AND POVERTY

INTRODUCTION AND BACKGROUND

Human population growth is sometimes held to underlie or even cause ecological problems. As will be seen below, there is reason to doubt this theory. Nevertheless there is also reason to believe that population cannot continue to grow indefinitely, and certainly not at the current rate, if the world is to become sustainable, or if basic human needs are to be satisfied. Hence population needs to be discussed in this work partly with a view to discovering its relation to ecological problems, whether global or local, and partly to considering the ethics of stabilising it at some sustainable level. The relation of population to food and to malnutrition will also be considered, in the course of relating population to global carrying capacity and to sustainable development.

Meanwhile poverty is a worldwide problem in itself, and is also widely held to comprise part of the explanation of some ecological problems. It may also be the underlying cause of population growth. Further, a sustainable world society can hardly coexist with poverty, either because poverty increases rates of fertility or because it is in any case inequitable and impractical to expect the poor to accept limits to their aspirations in the cause of sustainability. Consideration must accordingly be given to the question of whether, if this is the case, sustainability presupposes change away from poverty as a prerequisite of its own introduction.

A brief review of population statistics may serve to set the scene. The current global human population in 1998 is approaching six billion (where a billion is a thousand million), and has increased to this total from 2.5 billion in 1950. Growth is concentrated in Asia, Africa and Latin America. Rapid growth is continuing, although the annual rate of growth, which reached 1.9 per cent over the period from 1950 to 1985, is now reducing, not least in China. Short of apocalyptic disasters, growth is certain to continue because of the increasing numbers currently of child-bearing age or soon to attain it. If fertility reduces to replacement levels, and catastrophes are avoided, population will stabilise some half-century later; UN projections envisage stability (in the UN low scenario) at 7.7 billion in 2060, (in the medium scenario) at 10.2 billion in 2095; and (in the high scenario) at over 14 billion in the early part of the twenty-second century. Stabilisation would depend on a reduction of current growth rates down to replacement levels and on effective population policies, particularly in the Third World.[1] While UN projections are now treating the low scenario as seriously possible,[2] it would be imprudent to base any reliance or planning on a total of lower than 10 billion in the second half of the twenty-first century, although a lower target will be supported below. Thus the population of India seems set to outstrip during the early decades of the coming century that of China (currently 1.2 billion, and itself expected to reach 1.6 billion before stabilising).

Meanwhile in 1993, 1.3 billion people were living in extreme poverty, on less than the equivalent of one dollar a day. This figure represents some 23 per cent of the world's population. Some 70 per cent of these people are women. Even more people lack access to clean drinking water and sanitation, as was seen in Chapter 5. While definitions of poverty differ, and relative poverty can be a serious problem as well as the absolute poverty (poverty which precludes any tolerable quality of life) which is at issue here, this level of extreme (and absolute) poverty must be regarded as unacceptable. Recognising this, a series of United Nations Conventions and Resolutions aims to reduce this figure by 2015 to 0.9 billion, or 12 per cent of the projected world population of 6.5 billion.[3] To pragmatically minded people this target may seem ambitious; but in view of the uneven distribution of resources and the extent of unsatisfied basic needs which it appears to be prepared to tolerate, there is a strong case for regarding it as far too modest.

SOME PRO-NATALIST ARGUMENTS

Many ecological problems (nuclear fallout; acid rain and other forms of chemical pollution; holes in the ozone layer) are largely caused by high technology and, indirectly, by multinational corporations and international financial institutions (as has been seen in Chapters 5 and 6), rather than either by poverty or by population growth. Yet the growth in human numbers almost certainly contributes to other

UNIVERSITY OF GLAMORGAN PRIFYSGOL MORGANNWG

such problems, such as deforestation, desertification and global warming. Relatedly, the growing worldwide awareness of ecological problems has been accompanied in recent decades (as at the third UN Conference on Population and Development at Cairo in 1994) by an increased sense, particularly in the Third World, that policies of population planning are in the national as well as the global interest, and by an attenuation of pro-natalist claims, prominent at the first UN Population Conference at Bucharest in 1974, claims such as that population growth was desirable or necessary for development or for nation-building.

Interestingly, this kind of claim is sometimes grounded in a communitarian ethic which regards the addition of people likely to share one's own values and the enlargement of one's own community as outweighing the additional associated stress for other communities or for global ecosystems, whether from the increases thus advocated or from the greater increases which would ensue if everyone else did the same. Such disregard for non-members of one's community comprises a strong argument against the relevant forms of communitarianism, once the interests of all countries and their inhabitants are recognised as ethically relevant. Equally it forms a strong argument against forms of realism which seek to ground this kind of claim in national self-interest, with little or no regard for the global consequences. Communitarians can, however, support limits to population growth, albeit not (as cosmopolitans can) for the sake of humanity or of the biosphere; such support depends on limits to population growth being favoured currently among one's community, and would have to lapse if attitudes change (for example, through changes in religious affiliation), or even perhaps if changes of ideological fashion so dictate. Likewise realists can support such limits, provided that they can be reconciled with some suitable conception of self-interest; otherwise their support has to be retracted.

Yet cosmopolitan arguments for population growth are to be found. Julian Simon, for example, maintains that such growth enhances productivity and standard of living, and may be necessary to boost the process of development.[4] Simon does well to contest the claim of Paul Ehrlich that people actually comprise pollution,[5] a claim which in any case clashes with respect for persons and with consequentialist belief in the positive value of each worthwhile life, and he is probably correct also in holding that development in sparsely inhabited countries is harder to initiate than in others. But his case for continuing population growth is too optimistic about the effectiveness of free markets plus human resourcefulness in solving all significant problems, and pays insufficient consideration to social and environmental costs, and to the loss of species, habitats and biodiversity, certain to be associated with attempts to feed further billions of people.

A further cosmopolitan argument for population growth could be detected in Derek Parfit's Mere Addition Argument, which supports the 'repugnant conclusion' (derivable from consequentialism) that it is better, other things being equal, for there to be a population of many billions all leading lives which are worthwhile but barely so, than a smaller population all leading lives of higher quality.[6] However, this thought-experiment depends on all members of the larger population having lives of exactly the same (barely worthwhile) quality, with no inequality and no lives falling below the level at which life ceases to be worthwhile; and also on a like equality and absence of lives not worth living in the smaller population. Further, Mere Addition consists in the addition of extra people with no adverse effects on those already in existence. Parfit is, in fact, comparing entirely imaginary scenarios, not depicting empirical trends or practical policies. Hence it is possible to defend cosmopolitanism in the form of consequentialism, plus the positive value of each worthwhile life, and the 'repugnant conclusion' too, as I have done elsewhere,[7] without being committed in any way to additions to the total size of the current human population of our planet being desirable. The unmet basic needs of a billion of our contemporaries are a sufficient basis for quite different policies for the real world, since effort to meet that need is likely to make a greater positive difference than policies to increase the total could. Another ground for this view is the strong prospect that additions to the current population would add to the number of people with unsatisfied basic needs, while a third consists in the likely adverse impact of population growth on species and biodiversity.

Clearly pro-natalist arguments should be rejected for current circumstances. But before I discuss which policies best apply to population, the prospects for feeding an enlarged population and the relation of population growth to the environment need to be considered. For while facts and prospects cannot determine what should be done in the absence of values, they can draw to attention recognisable limits to morally acceptable policies.

FEEDING PEOPLE, PRESENT AND FUTURE

As David A. Crocker has remarked, it is important not to commit the 'fallacy of misplaced concreteness' or to focus solely on famines and their prevention, as opposed to the underlying problem of chronic malnutrition and the structures which cause it, or solely on producing enough food (which is compatible with people starving for lack of access) as opposed to the distribution of entitlements to food, the crucial factor where malnutrition is concerned. Besides, respecting people's characteristic capabilities involves not only provision for nutrition but also national and international development.[8] Crocker's points are well taken, and his stress on capabilities runs parallel with arguments in ethics for the realisation of essential capacities which I have presented elsewhere.[9]

Nevertheless, there may be upper limits to the population which our planet can sustainably support, or to the population which it can support without further loss of biodiversity, and such limits would be highly relevant to population ethics. Crocker recognises the relevance of food supply when he quotes research showing that 'since 1960 there has been sufficient food to feed all the world's people on a "near-vegetarian diet" and . . . "we are approaching a second threshold of improved diet sufficiency" in which 10 percent of everyone's diet could consist of animal products'.[10] But can this state of affairs persist?

Our Common Future summarises the relevant research on upper limits, which assumes that the area under food production remains close to the current level (1.5 billion hectares), and that yields rise from 2 tons of grain equivalent to 5 tons equivalent (highly questionable assumptions both). On this basis, the current consumption rate of calories per person suggests that potential production could sustain a little over 11 billion people, but only 7.5 billion if calorie consumption rises by 50 per cent (much less than is required for parity between the developed and developing countries). More could be fed if the area under food production could be increased and the productivity of areas of permanent pasturage could be enhanced sustainably.[11]

However, increases of the area under food production would intensify the problems of water supply, would press vulnerable marginal land into service, risking additional desertification, and would have an adverse impact on other species and on biodiversity. Genetic engineering, despite possible undesirable side-effects and the associated ethical problems involved in its deployment, might facilitate an increase of the productivity of permanent pasturage, but cannot be expected to solve the overall problem. In any case reliance on increased sustainable yields from the crop-growing land across the world to 250 per cent of current harvests is hazardous, and assumes large changes to food habits and to 'the efficiency of traditional agriculture',[12] possibly involving the forfeiture of the related traditional ways of life. Green Revolution agriculture has proved a mixed blessing, sometimes achieving efficiency in food production at the expense of environmental stress and of inequity among producers,[13] and excessive dependence on the continuation and extension of these methods could be counterproductive. Reliance even on the intactness of existing agricultural land and on existing fisheries is also problematic.

It is also notable that the possibility of feeding the population of the high scenario is not even considered; since feeding even the population of the medium scenario is highly problematic, the implication is that the high scenario does not amount to a serious possibility, as tens of millions would be likely to starve before such a high population could come about. If so, then international population planning should be premised on a ceiling of 10 billion, and aim at stabilisation somewhere below that level, as any higher population would be condemned to morally unacceptable levels of misery and premature death. A parallel conclusion is reached by Onora O'Neill on a basis of needs unmeetable if population grows excessively and of the requirements of a Kantian understanding of global justice.[14]

The points raised by Crocker effectively serve to lower the level at which the target for stabilisation should be set. For the possibility of producing a sufficient total of food for a given population does not guarantee adequate nutrition. Humanity has hitherto lamentably failed to distribute access to existing food to those who need it; and however much structures improve, plans should not be premised on total success in distributing access to food, or food entitlements,[15] to every human being. Since any failure of food distribution could involve vast suffering and starvation, and since it is in any case desirable for the average consumption of calories to increase, and this cannot happen if population growth prevents it, a significantly lower level of stabilisation should be aimed at than 10 billion, which should be seen as an extreme upper limit. (Further relevant considerations are introduced in the coming section.)

While the above arguments underline my earlier reply to pronatalist positions, they also strengthen the case for development as the only feasible alternative to widespread hunger. Without favouring all the solutions (such as the widespread use of high technology agriculture) proposed by developmentalists, we can consistently endorse both Amartya Sen's micro-economic explanations of famine, and also the kind of macro-conomic explanations supplied by Crocker for the underlying causes of malnutrition.[16] Both national and international development policies will be needed, as Crocker stresses, if the enterprise of feeding humanity through the coming century is to succeed.

POPULATION, POVERTY AND THE ENVIRONMENT

As previously mentioned, it is implausible that population growth causes the environmental problems associated with high technology. To suppose otherwise would be as unreasonable as blaming the Bhopal chemical disaster on its victims. Further examples of environmental problems with causes other than population growth or poverty include photochemical smog, radioactive fallout, oil slicks, acid precipitation and disruption of the ozone layer. However, population growth contributes to global warming, for poor people consume more energy in cooking a meal than those who can afford efficient cookers or microwaves (but much less per head per day than motorists). Concentrations of poor people are also prone to erode forests in their search for firewood (often the only fuel), and to pollute streams and rivers (often the only sewers), and often cannot afford to preserve resources like the soil by following tradition and allowing land to lie fallow.[17]

But as these connections reveal, population growth correlates closely with poverty, and even in these cases poverty contributes as much to environmental stress as population growth. Further, as Brundtland points out, poverty produces high rates of population growth, as children are needed 'first to work and later to sustain elderly parents', and the poor have no alternative resort.[18] Hence population strategies for poor countries need to tackle social and economic conditions as well as offering family-planning facilities. And strategies which do this actually do reduce fertility rates, or so WCED asserts and attests;[19] while Malthusians may deny this and assert that population invariably increases to the limit of food supplies, the evidence from the countries of Northern and Western Europe, Northern America, Australasia and Japan that fertility declines with prosperity (demographic transition) calls the credibility of Malthusianism into question.

While poverty and underdevelopment undoubtedly generate high rates of population growth, this growth contributes in turn to poverty. For increases in production are prone to be absorbed by the additional people, sometimes with few gains to quality of life becoming apparent (as is the case in many parts of Africa currently).[20] But poverty is a problem with many other causes, which need to be addressed alongside population growth. Meanwhile rapid population growth often proves to be incompatible with environmental integrity in much of the Third World,[21] although there is evidence to the contrary from the Machakos district of Kenya and elsewhere in Africa.[22] But as Fidel Castro points out, Third World nations, because of their lack of financial and technological resources, have no alternative but to overexploit natural resources, 'mismanagement' which 'engenders even greater poverty'.[23] While environmental problems have causes independent of population growth, many of the problems cannot be addressed without a reduction and eventual cessation of such growth. Radical policies to tackle poverty are necessary at the same time.

The correlation of population growth with poverty turns out to be an additional ground for population policies, since the alternative is the avoidable birth of large numbers of people whose basic needs are unlikely to be satisfied. For consequentialists, the foreseeable misery of possible people is a strong reason against bringing them into being. Such policies cannot be isolated from development policies (for reasons already given), which must be policies of sustainable development if obligations to future generations are to be met and if biodiversity and non-human species are to be preserved. For example, population growth rates often respond favourably to land reform, as Susan George points out, and sustainable policies of this kind can make a large positive contribution both to overcoming poverty and to limiting population growth, particularly in Latin America with its large and often uncultivated estates.[24]

The prospects for coming generations and also for the preservation of biodiversity strengthen the case not only for sustainable development, but also for population targets considerably lower than the maximum number of human beings who could in theory be fed. While each extra person has the potential for a worthwhile life and to contribute to the solving of problems, both local and global, even potentials must be understood as severely constrained when the very existence of their bearers conflicts with a healthy environment and with the continued existence of numerous other species. Additionally the value of non-human lives among species endangered by human population growth must count in favour of curtailing that growth. Thus, if ethical population policies can be devised, there is good reason, all things considered, to aim at a stabilisation of population at as early a date as possible, or in other words at the level of the UN low scenario (7.7 billion), or as near to that level as possible. The aim here proposed corresponds closely to the target adopted at Cairo of around eight billion.[25] The low scenario involves reaching replacement rates worldwide by 2010;[26] if that is in practice impossible, they should be reached as soon as possible thereafter.

POPULATION POLICY OBJECTIVES

A high degree of agreement was attained at the Cairo Conference on principles for population policies; and there are good grounds to accept most of these principles here, although with some significant reservations. Thus the sovereign right of each nation over the implementation of action on population is recognised in the opening paragraph of the Chapter on Principles, subject to respect for cultural diversity and for human rights;[27] such a framework must remain important as long as sovereign states remain, not least because in its absence little or nothing could be accomplished. But the inclusion of respect for human rights within this framework soon turns out to involve rejection of coercion, and thus implicitly of coercive national policies; all couples and individuals are held to have the basic right to decide freely and responsibly the number and spacing of their children and to have the information, education and means to do so (Principle 8).[28] While a general rejection of coercion is ethically sound, since coercion both undermines autonomy and is liable in any case to be counterproductive, the qualifications making the right to decide a right to decide responsibly, and to have the necessary education, could embody a partial concession to aspects of the Chinese population policy, which has travelled a long way down the road to coercion since the early 1980s. To that policy I will be returning; meanwhile the principle of a right to full information and to the means of implementing reproductive decisions (implicit in Principle 8) can readily be endorsed.

Another praiseworthy principle is the empowerment of women (Principle 4). Women's empowerment includes rights to education, to control over their own fertility, and to full participation in social and political decision-making.[29] Such empowerment promotes autonomy and personal fulfilment. At the same time the choices of educated women tend to contribute overall to lower birth-rates, to development and, arguably, to environmental conservation too.

Where birth-rates are concerned, it is likely that women currently give birth to more children than they want through lack of knowledge of and access to reliable contraception and safe abortion, and that access to education and to better services would allow their preferences to take effect much more than at present, not least in Third World countries. As Onora O'Neill draws to attention, the empowerment of women with regard to reproductive choices calls for a range of structural changes, likely to vary from context to context, transforming women's lives and opportunities;[30] and as Frances Moore Lappé and Rachel Shurman argue, it also involves replacing the subordination of women with much more democratic structures of decision-making, without which population growth and the ecological problems which relate to it may remain unsolved.[31]

A further principle concerns raising rates of child survival through improved health facilities, sanitation and security (Principle 11). Once again, this is something independently commendable; it is also held by developmentalists to be likely to contribute to couples having fewer children through an increased confidence that existing children will survive to adulthood. While Malthusians (including Malthusian environmentalists) might contest this reasoning, the various environmental pressure-groups represented at Cairo seem to have endorsed this element of the consensus on population policies.

The larger related principle concerns the need to integrate population policies with policies of development. Thus Principle 5 runs as follows: 'Population-related goals and policies are integral parts of cultural, economic and social development, the principal aim of which is to improve the quality of life of all people.' Even people who question the causal link between poverty and high birth-rates (despite all the historical evidence from demographic transitions) could accept this principal aim, together with the policies which conduce to it, and the need to integrate them with population policies. Those who recognise this link have stronger reasons to support such integrated policies because of the impact which may reasonably be expected from such policies on population growth, together with the favourable difference this in turn would make to population-related environmental problems.

However, two reservations should be entered about the Cairo agreement. The first concerns the lack of serious recognition of the need to restructure international financial and trading relations if sustainable development and the other measures discussed here are to be effective in the Third World. If international debt is allowed to remain, and if trading relations remain skewed against primary producers, as arguably they currently are, then too much poverty and inequality will remain for policies of women's empowerment or for integrated policies of sustainable development to become realities. In the immediate context of population levels, this could mean billions of births being unnecessarily generated by insecurity and by poverty. Special efforts in the fields of family planning and of perinatal health may bring significant reductions of population growth, but these reductions may well, in the absence of the further reductions attainable through structural change, prove inadequate, certainly if the UN low scenario is accepted as a target and perhaps even if we adopt targets based on the medium scenario. The Cairo Conference, despite its Principles, effectively paid too little attention to development objectives.[32]

Malthusian replies to all this might take the form of arguing that some countries have reached their carrying capacity, or else will reach it if their development is encouraged by aid, debt relief or restructuring of trading relations. Support for such countries leads only to inevitable collapse, at the stage where population eventually outstrips food supply (or water supply, or disease control); and as present support will worsen the future collapse, it is kinder (on consequentialist grounds) to withhold it. Here a response is in place, which involves distinguishing senses of 'carrying capacity', as William Aiken has done. A particular territory can have a determinate carrying capacity for a particular wild species, a capacity which cannot be exceeded without reducing the future capacity of that territory to support that species, but this strictly biological notion of carrying capacity does not apply to human beings, who can increase the carrying capacity of a particular country by practices such as aid and trade. For human beings, 'carrying capacity' is largely a socio-economic concept, and depends on factors like markets or the measures taken to regulate them. This being so, the claim that a country has exceeded its carrying capacity often presupposes a desire to protect the prosperity of more affluent places or to refuse to regulate markets or to render aid.[33] But this is a (short-sighted) appeal to self-interest, and not to the good of the vulnerable country, let alone to the greater good. Some kinds of regulation of international markets (as through the Kyoto agreement or through debt relief) can reasonably be held to be a requirement of sustainable development in the Third World, of enhancing quality of life in those countries, and of enhancing their long-term socio-economic carrying capacity as well. Hence practices of these sorts ought to be adopted, and on consequentialist grounds too, despite Malthusian claims to the contrary.[34]

THE CASE OF CHINA

A second reservation about the Cairo agreement concerns its stance on coercion, and its muted but implicit criticism of the population policy of China. Granted that coercion undermines autonomy and is normally counterproductive, should it be ruled out for all circumstances? Onora O'Neill, despite her affiliation to Kantian principles of respect for autonomy, recognises that if non-coercive measures fail there might be a case for 'lesser coercion' (justified coercion to limit future injustice and coercion) to curtail reckless procreation in 'emergencies', but not for coercion, for example, to protect property against those in need;[35] and consequentialists can readily endorse this stance on the basis of maximising the satisfaction of need. Might China actually be in this situation?

According to Chu-zhu Zhu, Director of the Centre for Population Policy at the University of Xi'an, the 1996 population of China is set to rise to 1.6 billion in 2035, and will peak at that level if the population policy succeeds. In the absence of such a policy, it would rise to around two billion by 2050. But this would be a catastrophe, as China, with over 20 per cent of the world's population, but only 7 per cent of the arable land of the planet, and less than 15 per cent of its own territory suited to agricultural use, cannot feed such a population, let alone do so sustainably. So birth-rates need to be further lowered. Resources are in any case needed for people with special needs or disabilities, and minorities (such as Muslims with their distinctive values) have to be partially exempted from the one-child policy which is necessary for the attainment of the overall target.[36] An implicit conclusion is that deterrents and compulsory abortions too are justified to deliver the policy, in addition to educational measures.

This conclusion seems to conflict with advocacy of increased decision-making for women and of increased democracy; but it could seriously be maintained that at local levels both these factors have increased within China, and that in the absence of current policy the large strides taken in the cause of development could be at risk of reversal. Manifestly it also conflicts fundamentally with Catholic opposition to birth control; but once again the cause of development would seem to justify a refusal to heed such opposition. There is, certainly, a case against the view that non-coercive measures have failed; for the total fertility rate (or average number of children per woman of child-bearing age) fell from 5.82 in 1970 to 2.72 in 1978, all before the coercive one-child-per-family policy was introduced.[37] While this rate had to be further reduced to below replacement level, it is not completely clear that non-coercive measures (as opposed to the compulsory one-child policy) would have failed to achieve this.[38] However, it is unclear that the policy of the 1970s, as interpreted in the different provinces, was significantly less coercive than the current policy, and this casts doubt on the evidence for the claim that non-coercive measures were sufficient and could have remained so.[39]

But there is also an ethical case against the current policy: even if circumstances require a limitation of reproductive freedom, this should be done in an impartial manner, which minimises harmful consequences,[40] particularly for the vulnerable, and it is not clear that these requirements have always been satisfied. Thus neither Kantians nor consequentialists are strictly obliged to endorse the Chinese population policy. On the other hand the serious consequences attaching to any failure of non-coercive measures in the period since 1978 suggest that considerably greater understanding should be shown towards a coercive policy than was shown at Cairo. It also suggests that situations in which coercion would be justified are not too distant from the real world, or in other words that coercive but impartial policies could be justified in readily imaginable circumstances.

The ethics of the Chinese population policy may seem far removed from the ethics of the global environment. In fact, however, at the very least the future of the entire natural environment of China is at stake. Given that the population of China comprises just over one fifth of humanity, and that Chinese practice could well set precedents for other Third World countries in the coming century, probably much more is at stake. In addition, the fact of Chinese economic growth means that the global environmental impact of China is bound to be considerably modified by its population policy. With respect at least to its population policy, the rest of humanity is heavily in China's debt.

OBLIGATIONS OF AFFLUENT COUNTRIES

Can population policies be confined to the Third World? While there are strong reasons of self-interest for most developing countries to adopt such policies, the more affluent countries cannot reasonably expect participation in international population agreements unless they too are willing to adopt policies limiting the environmental impact of their populations. There are, as I have argued in connection with global warming, many independent reasons for such restraint. But considerations of fairness can be seen to add to them.

The lifetime environmental impact of an average American has been estimated at thirty times than of the average Indian; such estimates may be of dubious accuracy, but can be accepted as rough approximations, supplying a credible comparison. If we also accept that average carbon emissions in India should rise to cater for basic needs, we should recognise that this cannot happen without disaster unless environmental impacts in the North begin to fall, with corresponding changes to Northern consumption and lifestyles. Meanwhile the co-operation of Southern countries in international action on population must depend on the availability of corresponding Northern co-operation, since the shared benefits are likely to uphold Northern prosperity (to some extent) as well as Southern survival and development. While Northern co-operation must include increased commitment to international goals of sustainable development and to the structural changes which these goals require, Southern commitment to such agreements is likely in the long-run to depend also on Northern restraint with regard to environmental impact. (The appeal to fairness thus turns out to be open to a consequentialist interpretation.)

While the environmental impact of Northern countries is more closely dependent on their technology than on population, population increases there still make a large difference. Some Northern countries have declining populations, and others have increases only through immigration, which must in some degree be permitted in a world dependent on international communication, training and trade. Nevertheless Northern countries ought in the light of the above considerations to be willing to adopt population targets compatible with the UN low scenario and with global population trends, and to adopt policies of family planning and education to make these targets capable of realisation at the same time as facilitating informed reproductive decision-making among their current populations. They should also be willing to adopt targets for maximum environmental impacts, both for carbon and for other damaging substances and processes, to prevent the continuation of globally disproportionate and unsustainable patterns of pollution and resource-usage.[41]

What grounds can be given for the adoption of such targets and policies? National self-interest goes some way towards grounding these obligations, particularly on a long-term interpretation; for countries have a strong interest in a sustainable future, likely to be unattainable without obligations of this kind. Thus realists and communitarians could accept these responsibilities in some degree; and communitarians would have stronger reasons where the values of their own community currently included the integrity of global ecosystems. But, granted that considerations of self-interest might also be adduced to seek exemptions from or reductions to the implicit burdens, it is important that there are strongly favourable cosmopolitan reasons too, grounded in the well-being of other people, of future generations and of other species.

These reasons include grounds of justice, the theme of Chapter 9; appeals, however, to international fairness as a basis for acceptance of globally beneficial agreements have been discussed above. They also include the alleviation of poverty and misery and the promotion of well-being, among human beings and other species of both the present and the future; while people are not always motivated to act on these bases, they cannot be disregarded as ethical considerations. But the only type of ethical theory which takes them seriously is cosmopolitanism, exemplified by the biocentric consequentialism defended here. Realism and communitarianism give these considerations a derivative and contingent recognition at most, dependent on whether they can be related either to self-interest or to current community values. Those who accept that they warrant direct recognition (whether self-interest or community values support them or not) are committed thereby to cosmopolitanism. In the context of issues of population and poverty, it is difficult to avoid this commitment.

NOTES

1. World Commission on Environment and Development (WCED), *Our Common Future*, ch. 4.
2. See David R. Francis, 'Global Crowd Control Starts to Take Effect', *Christian Science Monitor*, 89 (22 October 1997), 1, 9; summarised in *International Society for Environmental Ethics Newsletter*, 8.3, Fall 1997, p. 9.
3. Cmd.3789, *Eliminating World Poverty* (White Paper on International Development, 1997), 1.24, pp. 20f.
4. Julian L. Simon, *The Ultimate Resource*, pp. 240–75.
5. Paul R. Ehrlich, *The Population Bomb*.
6. Derek Parfit, *Reasons and Persons*, pt IV.
7. Robin Attfield, *Value, Obligation and Meta-Ethics*, ch. 10; *A Theory of Value and Obligation*, ch. 9.
8. David A. Crocker, 'Hunger, Capability, and Development', in William

Aiken and Hugh LaFollette (eds), *World Hunger and Morality* (2nd edn), 211–30.

9. Attfield, *Value, Obligation and Meta-Ethics*, chs 4–6; *Theory of Value and Obligation*, chs 3–5.
10. Crocker, 'Hunger, Capability, and Development', p. 217, quoting Robert W. Kates and Sara Millman, 'On Ending Hunger: The Lessons of History', in Lucile F. Newman (ed.), *Hunger in History: Food Shortage, Poverty and Deprivation*, p. 404.
11. WCED, *Our Common Future*, Box 4–1, pp. 98–9. For an optimistic account of prospects up to 2020, see Tim Dyson, *Population and Food*, ch. 7; for an optimistic appraisal of the possibilities of sustainable agriculture, see Jules Pretty, 'Feeding the World?', *Splice*, 4.6, August/September 1998, 4–6.
12. Ibid., p. 99.
13. George R. Lucas Jr, 'African Famine: New Economic and Ethical Perspectives', *Journal of Philosophy*, 87, 1990, 629–41, pp. 630–1; see also Vandana Shiva, *The Violence of the Green Revolution: Third World Agriculture, Ecology and Politics*, chs 2–5.
14. Onora O'Neill, *Faces of Hunger: An Essay on Poverty, Justice and Development*, p. 157.
15. Amartya Sen, *Poverty and Famines: An Essay on Entitlement and Deprivation.*
16. See William Aiken, 'Famine and Distribution', *Journal of Philosophy*, 87, 1990, 642–3.
17. Bread for the World Institute, *Hunger 1995: Causes of Hunger*.
18. WCED, *Our Common Future*, p. 106. See also Ben Jackson, *Poverty and the Planet*, pp. 181–2; and Frances Moore Lappé and Rachel Shurman, 'Taking Population Seriously', in Lori Gruen and Dale Jamieson (eds), *Reflecting on Nature: Readings in Environmental Philosophy*, 328–32, pp. 328–9.
19. WCED, *Our Common Future*, p. 106.
20. Bread for the World Institute, *Hunger 1995*, pp. 11–2.
21. See Holmes Rolston, 'Feeding People versus Saving Nature', in Aiken and LaFollette (eds), *World Hunger and Morality*, 244–63. By way of reply, see Robin Attfield, 'Saving Nature, Feeding People and Ethics', *Environmental Values*, 7, 1998, 291–304.
22. See Mary Tiffen, et al., *More People, Less Erosion: Environmental Recovery in Kenya*; Victoria Johnson and Robert Nurick, 'Behind the Headlines: the Ethics of the Population and Environment Debate', *International Affairs*, 71.3, 1995, 547–65, pp. 555–6.
23. Fidel Castro, *Tomorrow is Too Late*, p. 17.
24. Susan George, *How the Other Half Dies*.
25. See Peter G. Brown, 'Trusteeship and Consumption', unpublished paper presented to University of Maryland Conference on 'Consumption, Global Stewardship and the Good Life', 1994, p. 23.

26. WCED, *Our Common Future*, p. 102.
27. Program of Action, 1994 International Conference on Population and Development, *Population and Development Review*, 21, 1995, 187–220, ch. II: Principles, p. 190.
28. Ibid., Principle 8, p. 190.
29. Ibid., Principle 4, p. 190.
30. O'Neill, *Faces of Hunger*, pp. 157–8.
31. Lappé and Shurman, 'Taking Population Seriously', 328–32.
32. C. Alison McIntosh and Jason L. Finkle, 'The Cairo Conference on Population and Development: A New Paradigm?', *Population and Development Review*, 21.2, 1995, 223–60, p. 251; Neil Thomas, 'Who Defused the Population Bomb?', *Planet: The Welsh Internationalist*, 116, 1996, 85–92.
33. William Aiken, 'The "Carrying Capacity" Equivocation', in Aiken and LaFollette (eds), *World Hunger and Morality*, 16–25.
34. For a further consequentialist reply to neo-Malthusians, see Jesper Ryberg, 'Population and Third World Assistance: A comment on Hardin's Lifeboat Ethics', *Journal of Applied Philosophy*, 14.3, 1997, 207–19.
35. O'Neill, *Faces of Hunger*, p. 158.
36. Interview at Xi'an Technological University, July 1996.
37. Judith Banister, 'An Analysis of Recent Data on the Population of China', *Population and Development Review*, 10.1, 1984, 441–71, p. 454.
38. See Tyrene White, 'Two Kinds of Production: The Evolution of China's Family Planning Policy in the 1980s', in Jason L. Finkle and C. Alison McIntosh (eds), *The New Politics of Population: Conflict and Consensus in Family Planning*, 137–58.
39. For the information here I am grateful to my Cardiff colleague Neil Thomas, a specialist in population studies.
40. Thus Daniel Callahan, *Ethics and Population Limitation.*
41. I am grateful to William Aiken for access to an unpublished conference paper 'Development and Population Policy', which prompted several of the points conveyed here and elsewhere in this chapter.

CHAPTER 8

BIODIVERSITY AND PRESERVATION

Biodiversity was justifiably the subject of a Convention at the Earth Summit in Rio in 1992, despite intense political disagreements between the negotiators.[1] Losses to biodiversity, or biological diversity, have become so vast that the rate of loss may already be exceeding the rate of diversification implicit in the evolutionary process. Such biodiversity loss comprises a global problem, and in several senses at that. Thus it is worldwide in extent, and globally cumulative for particular types of habitat (such as wetlands); a significant fraction of particular global resources (such as rainforest species) is being lost through its impact in particular regions (such as habitat losses in Madagascar, Borneo and Brazil); and in the form of deforestation it is probably affecting global systems too, through the destruction of watersheds (for example, in Nepal) and the disruption of rainfall patterns and the systems which depend on them, and is thus multiplying global climate change.[2]

In this chapter, the nature and value of biodiversity will be investigated, and then the implications of its value for global policies of preservation and of sustainability, including issues surrounding the Rio Convention on Biodiversity (1992). Ethical issues concerning global strategies of preservation will also be discussed, as will the relative merits of some communitarian and cosmopolitan approaches to the protection of biodiversity.

WHAT IS BIODIVERSITY?

Biodiversity means variety and variability at the genetic, species and ecosystem levels, and includes diversity within individual species, diversity among species, and the diversity of their ecosystems and habitats; it is a matter not only of numbers of species or of ecosystems, but of the countless interconnections between them.[3] Yet numbers can be eloquent. According to United Nations Environment Programme (UNEP) estimates (which are admittedly controversial), there are about 30 million species on Earth, of which only about one and a half million have ever been described, and of which about a quarter risk extinction within the next 30 years.[4] Sub-species are endangered in like proportion, while the biotic associations and habitats at risk are innumerable.

There are, certainly, conceptual problems about the definition of a species,[5] comparable to the problems about identifying ecosystems, noted in Chapter 6 above. But working scientists assume that these problems are capable of being solved (species being, on one definition, populations whose members are capable of interbreeding and producing fertile offspring), and the same assumption will be made here. If so, then species are not merely subjective constructs of taxonomists, but are genuine units of the ongoing evolutionary process of speciation – human interventions permitting.

But this brings us back to the real problem. The majority of Earth's species, for example, are located in tropical forests;[6] but these forest habitats are disappearing at an inordinate rate. So are other prolific types of habitat, such as coral reefs; while yet others, such as coastal waters and estuaries, are being subjected to high levels of pollution worldwide. Beyond doubt, extinction rates are accelerating;[7] habitat destruction is so extensive that towards a million species could have been lost by the year 2000.[8] Climate change is set to be a major underlying factor, as whole biomes in the northern hemisphere (such as boreal forests) move polewards.

THE VALUE OF BIODIVERSITY

What values are at stake when biodiversity is lost? Diversity sometimes has aesthetic value, but is not plausibly valuable in itself, despite the age-old Principle of Plenitude[9] according to which the more diverse a world is the better. Many people, however, hold that the species or the ecosystems which comprise biodiversity have value independently of human interests, and this belief can be accepted by anyone who recognises the moral standing of non-human creatures such as farm animals and their wild relations. But the reason for this could be that species and ecosystems consist of individual creatures capable of flourishing and of being harmed, and not that species or ecosystems have moral standing over and above that of their members. Certainly the interests of species are ampler than the interests of their current membership, but this could be because of the interests of their future membership (actual or possible).

Indeed, granted that the interests of future members include the survival of vigorous species members in the present, there may be no more to the interests of a given species than the interests of the collectivity of its members, present and future; these interests include, of course, the continued existence of other species on which the given species depends. The value of a species could also be a function of the intrinsic value of its members, plus their individual or collective value for other species.[10]

Elliot Sober questions whether biodiversity has any value other than aesthetic value. Thus arguments for preserving endangered species because of unforeseeable future uses amount to arguments from ignorance; unless fortified by arguments about known probabilities, such arguments count for nothing. Slippery-slope arguments are inconclusive without grounds for holding that the slide is unstoppable. Arguments for preserving whatever is natural assume (implausibly) that whatever is natural is desirable. Claims that diversity must be preserved because it promotes stability are also based on what turns out to be a questionable assumption. Appeals to the good of wholes such as systems or species assume that such wholes have moral standing and an independent good, which is open to question (see above). And appeals to the rarity value of the last surviving members of a species either need supplementation or fail; it is implausible that value increases in inverse proportion to numbers.[11] It is also difficult to appeal both to rarity value and at the same time to the importance of preserving 30 million species in all their multiplicity.

The concern of environmentalists for preserving endangered species could have an ampler basis than rarity, or than Sober recognises. For the extinction of a species eliminates the lives of all the future members of that species, and these lives could have been valuable (intrinsically or otherwise). So too, admittedly, could have been the lives of members of another species which would colonise their niche; but members of the endangered species may have a distinctive range of capacities or a distinctive contribution to make to other species. And this already shows that there could be good grounds to preserve the last viable members of an endangered species in preference to the same number of members of a more plentiful species. The argument from the value of developing capacities would not impress people who seek to restrict moral standing either to intelligent or to sentient creatures (although the boundary of sentience already admits most vertebrates to moral standing); but this has been argued above (in Chapter 2) to be an arbitrary boundary, to which a biocentric position (recognising the moral standing or all living creatures) should be preferred.

But the wide moral constituency recognised by a biocentric ethic itself brings to attention the impossibility, stressed by John Passmore, of preserving everything, and the need for selectivity and for criteria to guide preservation policies.[12] Here it is worthwhile to review the arguments from human interests for preservation. Thus global food supplies depend on averting crop failures through the discovery of genes resistant to natural predators and carried by the wild relatives of food plants;[13] for example, the Brundtland Report cites a perennial variety of wild maize, found in a Mexican forest under threat of destruction, which could prove crucial to world production.[14] Again, almost a quarter of all medical prescriptions (or more, according to Brundtland) are for drugs extracted or derived from plants or from micro-organisms such as bacteria, and there is reason to expect research to yield many more.[15] Thus, in both the nutritional and the medical contexts the arguments are not based simply on ignorance, but on the likely existence of vital biological resources liable to be foregone if the relevant habitats are destroyed; and these arguments are strengthened by appeal to the value of wildlife-derived materials useful to industry.[16] Since, however, we cannot be certain which the relevant habitats are, there is a related precautionary argument against the destruction of habitats in general. This is not a conclusive argument for preserving all species, but it does support preserving habitats in proportion to the genetic diversity to be found in them. Habitats such as tropical forests and coral reefs would thus be high (but not unique) priorities for preservation.

The argument from resources can be summarised in the analogy of living nature as a genetic library. Destroying rainforest is comparable to burning a library 'of volumes that have not even been read'.[17] If this analogy related merely to arguments from ignorance, it would be inconclusive, for there could in theory be libraries which are worthless. But the preceding arguments from food, drugs and materials suffice to show that this is not a relevant fear; and the analogy further conveys the value likely to be derived from the study of nature, study unpredictable as to its outcomes, but likely to contribute significantly to the body of human understanding, and once again entirely dependent on policies of preservation; for at present, as Fidel Castro remarks, numerous species are being destroyed before they are discovered or can be studied.[18] This argument need not assume that knowledge is intrinsically valuable, as

Karen L. Borza and Dale Jamieson suggest,[19] or that its fruits are invariably (as opposed to generally) beneficial; a suitable alternative basis could be found in the intrinsic value of the pursuit of understanding. The view that opportunities for this pursuit should not be curtailed by the destruction of the world of nature is, as Borza and Jamieson affirm, a strong cultural argument in favour of preservation, albeit not a decisive one when the costs are also considered.[20]

Another analogy relates not to resources but to the dependence of humanity on nature: Anne and Paul Ehrlich's analogy of rivets. Populations and species are compared to the rivets which hold together an aeroplane; while many can be removed before the plane becomes unsafe, it is imprudent to rely on a plane from which rivets are regularly being removed.[21] But the systems on which humanity depends have wild populations and species as their rivets, and there is a finite limit to the amount of extinctions which they can tolerate. Analogies are not arguments, but if species were as vital to the systems on which humanity depends as rivets are to planes there would be ample reasons of prudence to halt extinctions. Analogies also have limits; for many species are too localised or too similar to others to have the role of a rivet, and in any case if the more rivet-like or crucial species could be identified, then other things being equal the rest could be discarded. The analogy does, however, bring out the facts that even from an anthropocentric perspective, nature consists of more than resources, and that it cannot be limitlessly damaged with impunity. A parallel anthropocentric argument for species-preservation is implicit in the preamble to the 1973 Convention on International Trade in Endangered Species of Wild Fauna and Flora (CITES); species are to be regarded as an irreplaceable part of vital natural systems.[22]

Quite a different ground for preservation has been advanced by Mark Sagoff, and could be extended so as to apply to biodiversity preservation.[23] This argument appeals to the cultural values of Americans, whose national experience was shaped by confrontation with nature, and for whom free-flowing rivers and species such as eagles symbolise freedom. It is also a communitarian argument, beset by the problems characteristic of such arguments; for even if these are the values of current Americans, other societies have different ideals and certainly different symbolism, and the same may apply to future American generations. But ecosystems do not respect national or cultural boundaries, and reasons for participating in their preservation will be required in every generation.[24] While cultural symbolism may be an important antidote to American reluctance towards international biodiversity programmes, it is no substitute for arguments of a more pervasive and perennial kind.

Granted that biodiversity loss largely results from deforestation and climate change, themselves associated much more with economic expansion and high technology rather than with population growth, an argument of Vandana Shiva for its preservation should be remarked. Biodiversity, she points out, has been a common resource (for food, medicines, fuel and housing materials among other uses) of local people in the Third World, who have preserved it up to the present; multinational corporations seek to exploit it, but are not reliant on it when profits can be otherwise generated. She also claims that systems of communal property recognise its intrinsic worth,[25] a generalisation supportable with reference to at least some indigenous cultures. To the extent that traditional medicine and other forms of local knowledge often rely on the intactness of local biodiversity, and are likely to remain thus reliant, this is an argument from the needs of the poor, and deserves to be taken seriously; while it is apparently another communitarian argument, it is grounded in the values of many societies and the needs of still more, and its appeal to justice between societies is implicitly cosmopolitan, and applies to agents universally. Social justice, then, supplies grounds for preservation of nature in Third World countries, and does not cease to do so when international co-operation would be required to bring it into effect. While developmental considerations will sometimes support actions involving biodiversity loss, it is important that the interests of the poor (and of development too) often uphold preservation.

While the previous argument appealed in part to the respect of some societies for the intrinsic value of nature in some of its forms, we should not forget the argument, already introduced during the above discussion of Sober, based on the intrinsic value of the flourishing of non-human kinds. This argument has a double bearing on biodiversity preservation. For those who recognise intrinsic value in sentient lives only, counting the prevention of animal suffering as morally relevant but unconcerned about the well-being of non-sentient nature, it has an indirect bearing, since, as Borza and Jamieson remark, this position involves an obligation to respect the habitats of wild animals, and thus to protect the species and sub-species of those habitats.[26] But for those who accept the biocentric view that the flourishing of all lives has intrinsic value in some degree or other, there are direct reasons for preserving wild species and the habitats and ecosystems on which they depend.

Often there will be good biocentric reasons countervailing the case for preservation, as when, for example, risks to the health of conscious or self-conscious creatures justify the destruction of pathogens; and sometimes, as when the interests of different species conflict, criteria compatible with biocentrism concerning, for example, the capacities of different creatures or their ecological relations will need to be invoked.[27] Unlike ecocentrism, biocentrism avoids making a (vulnerable) appeal to the supposed intrinsic value of the health of ecosystems, supporting ecosystem preservation rather through its importance to the well-being of creatures. A biocentric ethic supports biodiversity preservation directly, and also reinforces cultural arguments: aesthetic arguments turn out to concern preserving for appreciation, and scientific arguments to concern preserving for study, beings which are bearers of intrinsic value independently of study or appreciation. Everything which counts in favour of such an ethic discloses the incompleteness of the anthropocentric thesis toyed with by Passmore, that 'all preservation is cultural preservation' and that 'what is valuable is always human activity or human experience'.[28] Such anthropocentrism involves an excessively narrow view of ethics.

PROBLEMS ABOUT PRESERVATION

Policies of preservation, which (as has been seen) seek to maintain or restore a current or earlier state of affairs for the foreseeable future, are often contrasted with policies of conservation, which seek to protect resources with a view to their eventual use. The Rio Biodiversity Convention (1992) standardly adopts the anthropocentric usage of conservation, for example requiring each contracting party to adopt a programme of 'the conservation and sustainable use of biodiversity',[29] but is fortunately concerned with long-term protection and constraints on use as well. Policies of conservation treat nature as a stockpile of resources, but are prone to disregard the possibility of future benefits unforeseen at present, and the dependence of humanity on the systems of which the 'resources' form part (and thus the benefits of non-use), quite apart from their intrinsic value. Hence human interests (as well as the good of non-human kinds) often support long-term preservation rather than conservation, or, in the case of renewable systems, sustainable use of a kind compatible with their long-term intactness.

The case for preservation has its problems. Biological processes are essentially dynamic, and in the natural course of events involve episodes like forest fires, without which species would have evolved differently; hence only distorted forms of preservation seek to forestall all such episodes, as opposed to those which threaten human settlements or endangered species. For the same reasons, it is futile to attempt to preserve for all time one or another phase of a natural sequence, even though different species flourish in different phases, and thus not all can be preserved. There again, many species thrive only because of farming practices such as grazing; where such habitats have subsequently become overgrown, preservationists have a choice between restoring a close-cropped landscape and preserving either the new, post-grazing foliage or, what may be different again, the preceding ancient forest, and may defensibly choose modified nature rather than nature as it was before culture. In any case, many other values are often relevant besides those internal to preservation; if biodiversity is to be preserved partly for the sake of a blend of human and non-human interests, then the practice of preservation has to be consistent with the broader goal of sustainable development. However, a connection stronger than consistency is recognised in Europe; the European Union Habitats Directive (1992) treats the objective of 'maintenance of biodiversity' as contributory to the 'general objective of sustainable development'.[30]

There is also the issue of what general kinds of entity to preserve. Possible objects of preservation include sub-species, species and habitats, and these different projects generate different objectives, which could on occasion clash. A standard view is that preservation of habitats (such as the swamps of Florida) is liable to save a greater range of biodiversity than that of species (such as cougars) or of sub-species (such as the Florida panther). While this view will usually be justifiable, because of the range of individuals and of species preserved, there will also be exceptions, as where a particular habitat or a range of species cannot be saved but a species can, or where a sub-species is the only representative of its species in a particular region, which cherishes its survival there.[31] Diversity is, after all, a derivative value, and not the only value of relevance to preservation. The long-term prospects of the survival of whatever is preserved is also a relevant criterion, and its expected fecundity another. But granted our inability to identify most existent species and sub-species, and the very abundance of species in habitats like rainforests, the ocean bed and coral reefs, the preservation of particular habitats of these kinds will often be the unquestionable objective for the project of biodiversity preservation. Marginal cases will continue to offer scope for debate, but do not subvert the clarity of such goals as those of the preservation of habitats and of wildlife, goals which are crucial for a biocentric ethic. Nevertheless, integrating practices supportive both of these goals and of sustainable development is sometimes, as will be seen below, a different matter.

THE BIODIVERSITY CONVENTION (1992): SOME ETHICAL ASPECTS

Controversy about the Biodiversity Convention at the Rio conference focused neither on the value of biodiversity nor on the importance of preserving it for the general good of humanity, but on ownership of the products of its use in biotechnology.[32] While developing countries bear much of the cost of preservation, they are seldom able to reap the associated economic benefits. These benefits go to biotechnological industries, which market the commercial products of biodiversity for agricultural, pharmaceutical and comparable purposes. To do so they depend on access to biological diversity and frequently on the knowledge of local communities, often without payment to these communities or their countries. Often intellectual property rights (such as patents) are used to protect their products and techniques (as in the attempt by a US company in 1995 to patent drugs extracted from the neem tree, long-used for its medicinal and insecticidal properties in India),[33] and defended on the basis that these products are not wild species but derivatives, expensively isolated and synthesised.[34] Developing countries are aggrieved that the system by which the profits go to the corporations and the costs to themselves is inequitable.

Their case can be supported by considerations about the causes of biodiversity loss, which gives rise to the need for preservationist effort in the first place. As mentioned above, deforestation and climate change are prominent causes; and while climate change is now widely recognised to be anthropogenic and largely due to Northern production and consumption, deforestation and comparable forms of habitat loss are often due, as Vandana Shiva has argued, to large projects of mining, road-building and dam construction, or to the introduction of modern agriculture, fisheries and animal husbandry, often associated with Northern transnational corporations and international banks.[35] The facts that the governments of developing countries have co-operated in these enterprises and that developing countries have been among the intended beneficiaries does not make this irrelevant, in view of the lack of options for developing countries, the direction in which most of the profits have flowed, and the environmentally harmful misjudgements which have frequently been involved. The moral responsibility of developed countries for 'past policies of exploitation' has led Parvez Hassan to claim that they should pay their 'ecological debt' to developing countries, changing their trading patterns, transferring needed technology, funding biodiversity preservation and cancelling unpayable debts (in the more conventional sense of that term).[36] These issues extend beyond the scope of a chapter on biodiversity, and will be further considered in the next chapter, on global justice. But there is also a powerful ethical case here for greater equity in matters of international co-operation concerning biodiversity.

Biodiversity preservation, including programmes for the restoration of damaged areas and for co-operation with local people in the protection of wildlife, is expensive, and developing countries, while recognising its importance for sustainable development, understandably regard the eradication of poverty as a higher priority. Since most biodiversity preservation would be located in developing countries, these countries can reasonably ask wealthier countries which stand to benefit, which have destroyed most of their own forests, prairies and wetlands, and which have had some role in generating global environmental problems, to underwrite most of the costs. This view could be supported by the principle of equity that where common enterprises must be undertaken, those with the greatest unsatisfied needs should not be expected to make as large a contribution as the others. But other principles too are relevant. Since preservation involves developing countries foregoing some opportunities for development, wealthier countries seeking their co-operation, and also seeking to protect their own prosperity thereby, should be willing to compensate developing countries for opportunities foregone; the principle that compensation is due for opportunities foregone through participation in a shared undertaking can be accepted even by those reluctant to recognise the case for compensation for past exploitation. These principles of burden-sharing and of compensation are readily upheld by a consequentialist ethic (as well as by most other varieties of cosmopolitanism), and in the present case all the more so by consequentialism in a biocentric form.

Prior to the Rio Conference, the agreed international stance on biodiversity was that of the Food and Agriculture Organisation (FAO) International Undertaking on Plant Genetic Resources, which declared such resources a 'heritage of mankind', which 'should be available without restriction' and 'free of charge, on the basis of mutual exchange or mutually agreed terms';[37] and preparatory drafts of the Rio Biodiversity Convention had referred to biodiversity in general as a 'common heritage of mankind',[38] placing it on a par with the ozone layer, the atmosphere, and the minerals of the deep ocean bed (as in the Law of the Sea Treaty, which came into effect in 1994). But appeals to the common heritage of humankind, which could in other contexts be used to justify the international taxation of companies that extract minerals from international waters to fund the development of poverty-stricken inland states, and thus to reduce inequality, seemed likely in the case of resources located in Third World countries to have the reverse effect, of enriching transnational corporations at the expense of those countries. Other objections to 'common heritage' discourse are also in place; thus it represents biodiversity, and thus all the living constituents of the biosphere (apart from humanity), as ownable, and also as having value only in relation to human interests, positions shown to be unacceptable in previous chapters. The eventual preamble of the Convention speaks of biodiversity as a 'common concern of humankind' instead, an equally anthropocentric phrase,[39] but with less far-reaching political implications.

Developing countries, in any case, objected both to free access to the biodiversity of their territories and to the erosion of national sovereignty which the 'common heritage' principle could have involved. Thus the agreed objectives of the Rio Convention, expressed in Article 1, include 'the conservation of biological diversity, the sustainable use of its components and the fair and equitable sharing of the benefits', through 'appropriate access to genetic resources and appropriate transfer of relevant technologies . . . and appropriate funding'. This Article expresses the concerns of developing countries that the cause of development should be blended with that of preservation.[40] The Convention also incorporated Principle 21 of the Stockholm Declaration, which recognises that states have 'the sovereign right to exploit their own resources', subject to obligations not to damage 'the environment of other states' or of areas outside their jurisdiction, and to use their own resources sustainably. This conferred on states control of local resources and of access to them (and has required the FAO Undertaking to be revised).[41] While the principle of territorial sovereignty may some day need to be modified or replaced in the name of greater international co-operation, the likely effects of alternatives made its recognition at Rio well justified in current circumstances.

The Convention also recognised obligations for developed countries to provide financial assistance and technology-transfer to developing countries to enable them to fulfil environmental obligations alongside existing priorities (explicitly including the eradication of poverty), and for all countries to share equitably benefits arising from the knowledge and practices of indigenous and local communities with the communities concerned.[42] But it did not include the notion of 'ecological debt' for past exploitation, which was successfully resisted by Northern countries,[43] nor a list of globally important areas and species for special protection, successfully resisted by developing countries fearful that it could be used to curtail development.[44]

The only government participating in the Rio Conference that refused to sign the Convention was the US Bush administration (a decision reversed a year later by the Clinton administration). Their objections alleged insufficient recognition of intellectual property rights and excessive control in the hands of developing countries over the Global Environment Facility (the international fund set up in 1991 and endorsed by the Convention); but the US delegation often found itself in a minority of one, and deserted by customary allies such as Japan and the countries of Northern Europe.[45] Debates about these matters diverted attention from pro-active environmental policies; hence informed commentators such as Mustafa Tolba, executive director of UNEP, judged the Convention as representing 'the minimum on which the international community can agree'.[46] In the subsequent period, USA has effectively reversed the Rio position on control of access to biodiversity through agreements reached on trade-related intellectual property rights during the Uruguay Round of the General Agreement on Tariffs and Trade (GATT) (1993), and has put strong pressure on numerous governments to recognise such intellectual property rights. This pursuit of national advantage amounts to a retreat from the international equity principles of Rio, and would be difficult to defend on the basis of any ethic other than realism. Meanwhile effective control of the Global Environment Facility has been secured by the donors from the North, and its funding is far from adequate.[47]

While a sequence of Conferences of the Parties has increased the size of the world's protected natural areas, further consideration should be given to Brundtland's somewhat modest proposals for the total expanse of protected areas to be tripled (as compared with just over 4 million square kilometres of 1987), so as to preserve a representative sample of the Earth's ecosystems,[48] for sufficient international funding to be made available to facilitate this on a sustainable basis, and at the same time for the causes of environmental degradation to be addressed. There should also be a worldwide policy of involving local people in wildlife preservation, rather than a continuation of the prevalent policy of excluding them from protected areas (a counterproductive policy which benefits neither local people nor wildlife); and institutions should be devised for the equitable sharing of benefits arising from local knowledge with the communities concerned, as envisaged in Article 8 of the Convention, and in the preamble on indigenous and local communities.[49]

The Convention itself is so frequently qualified by phrases like 'equitably' and 'as appropriate' that Yamin commends an analysis by philosophers offering 'explication and choices for future interpretation and development by negotiators'.[50] That task cannot be undertaken here, and must be left for others. She also commends the development of an ethic concerned with global justice which would be intergenerational and non-anthropocentric as well as international, and which could then prove to be a resource for interpreting and developing international conventions.[51] This book and this series could be seen as initial contributions towards that large task.

STRATEGIES OF PRESERVATION

Besides issues concerning international relations, the goal of biodiversity preservation raises further global ethical issues. One issue which should be raised concerns whether preservationist policies are always justified by the goal of long-term preservation. It has already been mentioned that the goal of conservation (human benefit) may fail to be met by policies of conservation. However, David Schmidtz has persuasively argued that preservationism too can prove to be self-undermining. Hard-line preservationist laws and systems prohibiting game-hunting, such as policies of shooting poachers on sight, can be shown in at least some cases to be counterproductive as well as inhumane, and sometimes more wildlife would be preserved by introducing or tolerating systems which allow limited game-hunting and at the same time motivate local people through their receiving a share of the benefits.[52] As Brian Child has argued, 'wildlife will survive in Africa only where it can compete financially for space. The real threat to wildlife is poverty, not poaching'.[53] Local people seldom seek to decimate the wildlife of their own environment, but understandably give priority to making a livelihood. Systems are needed which allow of livelihoods through rewarding wildlife protection (for example for the sake of fostering tourism) rather than through poaching or through expanding the rearing of cattle in effective competition with wildlife.[54]

Thus conservationist (or 'wise use') policies sometimes (but far from always) promote the goals of preservation better than preservationism, as both libertarians and developmentalists are free to recognise. Considerations of animal welfare will sometimes count against such policies, where, for example, these policies would involve culling elephants, although sometimes they reinforce bans on hunting grounded in species preservation, as with the 1986 moratorium on whaling of the International Convention for the Regulation of Whaling (ICRW). But in cases where more animals would survive through such conservationist policies than in their absence, the bearing of these considerations becomes ambivalent.

As Schmidtz points out, attempting to put consequentialist principles, such as principles of preservation, into institutional practice 'can be a bad idea by the lights of the principle itself', at any rate where doing so will not make things better by the same lights; and 'whether doing so makes things better depends on circumstances'.[55] Or rather, institutional practices apparently geared to promoting preservationist goals may fail to deliver them. Consequentialism, of course, counts a broader range of values as goods than the survival of individual animals, or (come to that) than the preservation of biodiversity (goods including, for example, sustainable development), and this is a further ground (in addition to the gaps between intention and accomplishment and between appearance and reality) on which consequentialists can recognise situations in which practices intended to produce particular goods fail to produce those goods or to make things better overall. They are committed rather to supporting practices which, considered overall, deliver an optimal balance of value and disvalue;[56] and circumstances may show these not to be practices of hard-line preservationism, or even of overt preservationism at all. This, however, is not an argument for moral scepticism, but an argument against doctrinaire solutions, and in favour of taking the causes (including the social causes) of wildlife depletion seriously when devising policies of preservation.

To consider another aspect of ethical preservationism, biodiversity preservation should be understood as including cultivars, the strains of crops produced by human selection, and not least landraces, or indigenous varieties of cultivars, as well as natural species and subspecies. Landraces form part of the knowledge of Third World farmers, and are just as likely to be needed in the future as the wild counterparts of crops such as maize, wheat or potatoes. The biotechnology industry is prone to seek a monopoly for its own varieties, not only discouraging the cultivation of landraces, but tending to eradicate them altogether.[57] This means that the seed-banks in which many have been collected (such as the various Vavilov centres) should be regarded as a high priority for preservationist support and funding. The same often applies to the knowledge of local communities; the Biodiversity Convention recognised this,[58] but much remains to be done to implement this recognition before such knowledge is lost.

Finally, the kind of preservationist strategy should be mentioned in which the debt of a developing country is cancelled in exchange for the designation of a biodiversity-rich section of its territory as a protected area, thenceforth to be left undeveloped ('debt-for-nature swaps'). The main beneficiary during the 1980s was Costa Rica, but its total debt was only reduced by 2 per cent. Such deals have major advantages, including biodiversity preservation and the facilitating of development for a country liberated from the burden of debt, and as such were at one time welcomed, among others, by the prominent developmentalist Susan George.[59] But they also have significant disadvantages, including potential loss of sovereignty on the part of the debtor country, and the foreclosure of its access to the natural resources (including sometimes to the biological resources) of the designated area.[60] They are also typically slanted towards the interests of the donor, and amount to a piecemeal form of neo-colonialism, not too far different from a reintroduction of colonialism, old-style. While this may not discredit them for all circumstances (as sometimes all the other options will be worse), a better approach would consist in a less restrictive form of debt cancellation, involving more technology-transfer and less conditionality. There is a strong case for debt cancellation independent of the project of preservationism, and cancellation packages involving consultation with a debtor country over, and support for, its developmental and environmental priorities (rather than expropriation of pockets of territory) would be both more likely to generate the active support of its citizens and more respectful of its sovereignty and of the autonomy of its people. Such cancellation packages might also be more effective in terms of long-term preservation; for externally imposed restrictions on resource-use will not last forever in poor and unstable countries.

PRESERVATION: BIOREGIONAL OR GLOBAL?

One kind of communitarianism suggests that people can normally only be motivated to care for and preserve the environment of their locality, and that the factor of scale makes global environmental problems incomprehensible. Hence society should be organised on the basis of natural territorial units with which people can identify (bioregions like the catchment area of a river, such as the River Dart in Devon), and efforts should be made to promote the local self-sufficiency of such regions. Global concern can only arise on the back of a well-entrenched sense of local or bioregional identity. Such is, approximately, the position of bioregionalism.[61] The less attractive features of globalisation add to the attractions of this advocacy of life organised on a human scale.

Praise is in place for the fostering of a sense of place and of local identity, and of close acquaintance with the particularities of the natural and cultural history of a person's environment, whether native or adopted. Decisions should indeed be taken at as local a level as possible, as the European Union principle of subsidiarity enjoins. An exclusive preoccupation with global issues would, paradoxically, signify a narrowing of personality. But the same applies to a preoccupation with local particularities; and a region which deliberately sought economic and intellectual self-sufficiency would cut itself off from experience of cultural as well as biological diversity, and from patterns of fulfilment for its inhabitants, as well as from its ties and responsibilities to others. It would also be itself a cultural construction, for regions are no more natural to human society than local carrying capacities have been seen to be (see Chapter 6). Much to be preferred is the principle that all local communities should be respected by every such community, as recently commended by Bryan G. Norton and Bruce Hannon. This is, of course, a cosmopolitan rather than communitarian principle,[62] albeit one which, when harnessed to the view that central decisions should be minimised, pays too little heed to global problems and to the need for concerted global action in their regard.

Certainly if bioregionalism became the planetary pattern, global issues would remain untackled, soon making most if not all localities unsustainable. But this is only one reason why global concern is indispensable, including concern for global biodiversity preservation. For even if the moral isolation of communities was possible in centuries past, it is possible no longer. Moral ties with distant peoples exist, cannot be denied, and are daily re-emphasised by the media, even when the same media seek to deny them. Happily concern for places distant in space and time does not wax and wane with intensity of sense of local identity. Meanwhile culture, trade and travel continually blur national (let alone bioregional) boundaries, supplying participants with networks of relationships without depriving them of roots. Concern for biodiversity preservation, then, need not depend on a sense of local identity, much as this can nourish such concern; it has many springs, some more local and some more global, and is not confined, even as to its origins, to local watersheds or to their guardian spirits.

NOTES

1. Abby Munson, 'The United Nations Convention on Biological Diversity', in John Kirkby et al., (eds), *The Earthscan Reader in Sustainable Development.*
2. Open University, 'Biodiversity', in pt IV of *Environmental Ethics* (Course T861), p. 3.
3. Kirkby et al., *Earthscan Reader*, p. 15; Lori Gruen and Dale Jamieson (eds), *Reflecting on Nature*, p. 333.
4. UNEP, 'The State of the Global Environment', *Our Planet*, 4.2, 1992, 4–9; Munson, 'The United Nations Convention', p. 55.
5. See Holmes Rolston III, 'Duties to Endangered Species', in Holmes Rolston III, *Philosophy Gone Wild: Essays in Environmental Ethics*, 206–20, pp. 209–10.
6. Anne H. Ehrlich and Paul Ehrlich, 'Extinction: Life in Peril', in Gruen and Jamieson, *Reflecting on Nature*, 335–42, p. 339.
7. Martyn Murray, 'The Value of Biodiversity', in Kirkby et al., *Earthscan Reader*, 17–29, p. 19.
8. Norman Myers (ed.), *The Gaia Atlas of Planet Management*, p. 154.
9. Arthur O. Lovejoy, *The Great Chain of Being*.
10. See further Robin Attfield, *The Ethics of Environmental Concern*, ch. 8; *Value, Obligation and Meta-Ethics*, ch. 2.
11. Elliot Sober, 'Philosophical Problems for Environmentalism', in B. G. Norton (ed.), *The Preservation of Species*, 173–94.
12. John Passmore, 'The Preservationist Syndrome', *Journal of Political Philosophy*, 3.1, 1995, 1–22.
13. Murray, 'Value of Biodiversity', p. 22.
14. WCED, *Our Common Future*, p. 155.
15. Ibid.
16. Ibid., pp. 155f.
17. Murray, 'Value of Biodiversity', p. 25.
18. Fidel Castro, *Tomorrow is Too Late*, p. 31.
19. Karen L. Borza and Dale Jamieson, *Global Change and Biodiversity Loss: Some Impediments to Response*, pp. 10–11.
20. Ibid., p. 11.
21. Ehrlich and Ehrlich, 'Extinction: Life in Peril', p. 335.

22. CITES (Convention on International Trade in Endangered Species of Wild Fauna and Flora, 1973), *Text of the Convention.*
23. Mark Sagoff, 'On Preserving the Natural Environment', *Yale Law Journal*, 84, 1974, 205–67.
24. Borza and Jamieson, *Global Change and Biodiversity Loss*, pp. 9–10.
25. Vandana Shiva, *Biopiracy: The Plunder of Nature and Knowledge*, pp. 66–9.
26. Borza and Jamieson, *Global Change and Biodiversity Loss*, p. 12.
27. See further Attfield, *Value, Obligation and Meta-Ethics*, ch. 6.
28. Passmore, 'The Preservationist Syndrome', p. 20.
29. Biodiversity Convention, Article 6; Munson, 'The United Nations Convention', p. 57.
30. Habitats Directive (92/43/EEC), as quoted in Chris Miller, 'Attributing "Priority" to Habitats', *Environmental Values*, 6.3, 1997, 341–55, p. 349.
31. K.S. Shrader-Frechette and E. D. McCoy, 'Biodiversity, Biological Uncertainty, and Setting Conservation Priorities', *Biology and Philosophy*, 9.2, 1994, 167–95.
32. Murray, 'Value of Biodiversity', p. 21.
33. Farhana Yamin, 'Biodiversity, Ethics and International Law', *International Affairs*, 71.3, 1995, 529–46, p. 538.
34. Murray, 'Value of Biodiversity', p. 25.
35. Vandana Shiva, *Monocultures of the Mind: Perspectives on Biodiversity and Biotechnology*, p. 68.
36. Parvez Hassan, 'Moving Towards a Just International Environmental Law', in Simone Bilderbeek (ed.), *Biodiversity and International Law*, p. 75. See also Castro, *Tomorrow is Too Late*, pp. 19–24.
37. International Undertaking on Plant Genetic Resources, FAO Resolution 8/83; quoted by Yamin, at p. 541.
38. Yamin, 'Biodiversity, Ethics and International Law', p. 540.
39. Ibid.
40. Biodiversity Convention, Article 1; Yamin, 'Biodiversity, Ethics and International Law', p. 540.
41. Biodiversity Convention, Articles 3 and 15; Munson, 'The United Nations Convention', p. 56; Yamin, 'Biodiversity, Ethics and International Law', p. 541.
42. Biodiversity Convention, Articles 20 and 8; Yamin, 'Biodiversity, Ethics and International Law', p. 541.
43. Yamin, 'Biodiversity, Ethics and International Law', p. 542.
44. Munson, 'The United Nations Convention', p. 61; Yamin, 'Biodiversity, Ethics and International Law', p. 544.
45. See Lisa H. Newton and Catherine K. Dillingham, *Watersheds: Classic Cases in Environmental Ethics*, ch. 9, 'North Against South: The UNCED Summit at Rio de Janeiro'.
46. Munson, 'The United Nations Convention', p. 61.
47. Yamin, 'Biodiversity, Ethics and International Law', p. 542; see also

Joyeeta Gupta, 'The Global Environmental Facility in its North-South Context', *Environmental Politics*, 4.1, 1995, 19–43.

48. WCED, *Our Common Future*, pp. 165f.
49. Yamin, 'Biodiversity, Ethics and International Law', p. 541.
50. Ibid., p. 543.
51. Ibid., p. 546.
52. David Schmidtz, 'Why Preservationism Doesn't Preserve', *Environmental Values*, 6.3, 1997, 327–39.
53. Brian Child, 'The Elephant as a Natural Resource', *Wildlife Conservation*, 1993, 60f.; quoted in Schmidtz, 'Why Preservationism Doesn't Preserve', pp. 331–2. See also, in connection with India as well as Africa, Ramachandra Guha, 'The Authoritarian Biologist and the Arrogance of Anti-Humanism', *The Ecologist*, 27.1, 1997, 14–19, and his Postscript to 'Radical American Environmentalism and Wilderness Preservation: a Third World Critique', in Ramachandra Guha and Juan Martinez-Alier, *Varieties of Environmentalism: Essays North and South*, pp. 102–8.
54. Schmidtz, 'Why Preservationism Doesn't Preserve', pp. 331–5.
55. Ibid., p. 337.
56. See Attfield, *Value, Obligation and Meta-Ethics*, chs 7–11.
57. Shiva, *Biopiracy*, p. 51.
58. Biodiversity Convention, Article 8; Yamin, 'Biodiversity, Ethics and International Law', p. 541.
59. Susan George, *A Fate Worse than Debt*, p. 168.
60. Castro, *Tomorrow is Too Late*, pp. 38–9.
61. See Kirkpatrick Sale, *Dwellers in the Land: The Bioregional Vision*; Doug Abberley (ed.), *Boundaries of Home: Mapping for Local Empowerment*.
62. Bryan G. Norton and Bruce Hannon, 'Environmental Values: A Place-Based Approach', *Environmental Ethics*, 19.3, 1997, 227–45, p. 244.

PART III
GLOBAL JUSTICE AND GLOBAL CITIZENSHIP

CHAPTER 9

ENVIRONMENTAL JUSTICE AND WORLD ORDER

In Part Three, global justice and global citizenship are explored. Can a theory of justice be presented capable of accommodating both sustainability and development, and also the issues arising from the above discussions of human survival, climate change, resources, population and biodiversity? In attempting (in this chapter) to sketch such a theory, I will first consider the scope of justice, and then obligations to future generations and the suggestion that provision for these generations will deliver all the objectives of environmental concern. Justice between members of the current generation, and justice between species will also be considered, as is appropriate in a work on global ethics that defends a cosmopolitan view of the scope of moral standing. Some implications of the emerging theory for systems of property, for national sovereignty and for international relations will also be considered. Chapter 10 will revisit sustainability with particular reference to principles of intergenerational equity, while the final chapter will sift relevant conceptions of global citizenship.

THE SCOPE OF ENVIRONMENTAL JUSTICE

Environmental justice extends at least as far as environmental rights do; and as mentioned in Chapter 2, cosmopolitans (including those of the consequentialist and biocentric persuasions) are free to appeal to human rights, and to animal rights too, such as the right to a decent environment. As Principle 1 of the Rio Declaration states, 'Human beings . . . are entitled to a healthy and productive life in harmony with nature'.[1] These entitlements involve corresponding duties, sometimes attaching to countries, multinationals or international bodies; such duties should be understood as subject to Principle 3: 'The right to development must be fulfilled so as to equitably meet developmental and environmental needs of present and future generations.'[2]

However, obligations with regard to future generations go beyond the realm of rights, for (as explained in Chapter 2) the identity of members of future generations (except those already conceived) is not yet determined, and so no particular future beings (with the same exception) have rights against current agents. Nevertheless whichever ones eventually live will be affected for better or worse by current actions; hence we have in their regard what Kant called 'duties of imperfect obligation', duties not owed to assignable individuals but no less real than duties corresponding to assignable rights. As they will also benefit or suffer from distributions of goods and evils partly determined in the present, it is reasonable to regard these duties, with Brian Barry,[3] as obligations of intergenerational equity or justice. To sell future generations short is both wrong and inequitable. So too is a preoccupation with justice between generations at the cost of a neglect of justice between contemporaries.

While obligatory actions are ones which it is wrong not to perform, not all wrong actions are infractions of justice. Obligations to family and friends, for example, are not considered obligations of justice or equity, perhaps because they usually lack a background of conflicts of interest. Much less do obligations of justice exhaust the full extent of morality, much of which is not concerned with obligations at all, but with supererogatory traits such as kindness and generosity, which are morally admirable without being obligatory. Theories of morality are defective unless they provide for supererogatory acts and traits, and also for dispositions such as virtues, which are sometimes exercised in discharging duties (courage, integrity, conscientiousness) and sometimes take agents beyond the call of duty (to acts of compassion or mercy).

Critics of accounts of morality for which obligation and principles are central sometimes propose instead an ethic of care,[4] sometimes associating this with an Aristotelian theory of ethics for which character is more fundamental than right or wrong conduct (virtue ethics). Virtues are indeed indispensable for the moral life, but the concept of virtue is unsuited to supplying adequate guidance in decision-making contexts, particularly ones involving contemporary ethical conflicts or dilemmas. Without principles of obligation, virtues are short-sighted if not blind. This is amply illustrated by environmental issues concerning choices of what habitats to preserve, what technology to use, or what limits to pollution to require, as by other ethical issues such as ones of punishment or welfare policies. So theories about virtues or about caring cannot supersede or supplant theories of obligation or of justice; at the same time this fact does not make virtues and caring any less important in matters of practice.

Besides obligations towards future generations, other duties of imperfect obligation concern duties towards non-human animals and (on a biocentric view) other living creatures. Animal suffering is widely regarded as morally significant, a recognition which simultaneously acknowledges the significance of animal well-being, and also of human obligations with regard to animals, and which distances all who hold it from anthropocentrism. Believers in animal rights have a reason as such to regard these obligations as obligations of justice. But even people who accept obligations with regard to animals without accepting animal rights can regard these obligations as a matter of justice, concerned with the satisfaction of needs and interests which potentially clash with others; such obligations do not have to be owed to individuals, but hold with regard to whatever creatures may populate (say) given zones or regions. For example, refraining from depriving wild animals of the habitats in which they have evolved can be understood as an obligation of this kind. On a biocentric view, such obligations extend beyond animals to all creatures with a good of their own (all living creatures, that is), and may be obligations of justice on the same basis as obligations towards animals (although biocentrists are not obliged to draw this conclusion). Environmental justice, then, plausibly extends at least to the whole animal realm, and possibly beyond that to the full range of bearers of moral standing, as well as to future generations.

JUSTICE BETWEEN GENERATIONS AND SPECIES

Here it may help to recapitulate the principle concerning equity between generations adopted in Chapter 5 (on the basis of the consequentialist approach introduced in Chapter 2). Current agents, to the extent that they have the necessary powers and resources, have obligations to provide for the satisfaction of the basic needs of future generations, and to facilitate the development in the future of characteristic human capacities, and of the characteristic capacities of other species, to the extent that such satisfactions and development can foreseeably be facilitated. Two provisos were added: first, the condition that basic needs of the present matter as much as like needs arising in the future, and generate comparable (and potentially conflicting) obligations; and second, the further condition that future-related obligations hold only where factors beyond present control (such as climate change or future decisions) are not likely to prevent these good states of affairs from coming about. The second proviso limits these obligations to delivering what is within the control of current agents; in practice, doing this would often be restricted to facilitating opportunities for the satisfaction of future needs, or to avoiding outcomes (such as nuclear pollution) which could prevent such needs being satisfied.

The first of these provisos is intended to ensure that equity among contemporaries (such as the people of the present), or intragenerational equity, is not marginalised through emphasising equity between generations (intergenerational equity). Some environmentalists are tempted to ignore the needs of poor or vulnerable people of the present, while others (such as the Environmental Justice Movement and its members) prioritise these environmental and other needs,[5] adding a less emphatic recognition of the needs of the future. Here I have already argued that there are genuine obligations towards future generations (see above and Chapter 2), on the basis that needs should not go unsatisfied where agents can contribute to satisfying them. But this same basis supports obligations to satisfy the needs of poor or vulnerable contemporaries, as I have argued elsewhere.[6] Hence an inconsistency is involved where theorists focus either on current needs or on future needs exclusively.

However, the principle about future generations is a principle of equity, concerning, as it does, distributions of goods and evils between the holders of potentially conflicting interests. As such, it can be compared with two recent accounts of environmental justice, one explicit and one implicit. The explicit account is presented by Peter Wenz in *Environmental Justice*, and the implicit one is that of Bryan G. Norton in *Toward Unity Among Environmentalists*, which explicitly focuses on obligations to the future. In presenting a general theory of environmental justice, Wenz adopts a largely similar account of moral standing to the one presented here, adding a concentric account of moral priorities, in which family and compatriots form inner circles and come before foreigners, future people, members of other species (which have negative rights only) and evolutionary processes.[7] (While I do not regard evolutionary processes as having moral standing independently of their living products, that is not the key issue here.)

A positive feature of Wenz's approach is his recognition that obligations arise 'when someone is in need and someone else is in a unique position to be helpful',[8] although the view adopted in the present work does not restrict obligations to those whose position is unique in this way. But the problem about his position as a theory of justice is that the moral importance of a need is made to depend on which circle its holder inhabits, and thus on the strength of relationship between the holder and the subject of an obligation; future people, for example, belong to 'a relatively remote concentric circle'.[9] This feature, however, discards the basic principle of equal consideration (which is also an assumption of consequentialism), namely that like interests count alike. (I return to the meaning of this principle below.) Certainly there are special obligations which require a person to give particular consideration to members of her or his immediate family; but good general grounds can be given for preferring social systems involving families and expectations such as these over systems which do not, grounds consistent with equality of consideration overall, and which appeal to the good of all alike. To adopt instead, as basic principles of justice, principles which prioritise members of certain circles as such over members of others is to stray from recognisable pathways of impartiality and of equity.

Wenz's theory seems to account for the limited nature of human obligations in the matter of intervention to save animals from pain and predation.[10] But such interventionism would generally be so counterproductive with regard to the flourishing both of predators and of prey, and so disruptive of evolutionary processes, that biocentric consequentialism can readily justify non-intervention as a general stance, punctuated by intervention in special circumstances, for example to save a species or sub-species from extinction. Meanwhile a theory which removes the positive interests of wild animals from consideration altogether (such as their need for the kind of habitat in which creatures of their kind have evolved) is arbitrary, and serves to discard factors such as their well-being which cannot equitably be disregarded.

Since wild animals have moral standing, interests of theirs matter, morally speaking, on a par with the like interests of other creatures with moral standing. As I have argued elsewhere, following Donald VanDeVeer, interests can be alike as to their importance in the life of the creature concerned, and there again with regard to the significance of the psychological capacities which they involve.[11] Thus where greater interests, such as interests in self-consciousness or autonomy, are endangered and conflict with lesser interests, they take priority, but where psychologically significant capacities are not at stake, the basic interests of creatures count alike. (This departure from VanDeVeer's conclusions is also explained elsewhere.)[12] The principle of treating like cases alike irrespective of species boundaries is an implication of the cosmopolitan and biocentric approach to ethics. This is also the kind of theory for which Farhana Yamin recognises a need.[13] Paul Taylor too has presented a consequentialist and biocentric theory of ethics;[14] but the consequentialist theory presented here has a wider scope and its principles arguably have greater resources for supplying guidance for practical decision-making.

Norton's position is openly focused on practical effectiveness as much as on considerations of theory. He maintains that discharging obligations to future human generations will deliver at the same time the full range of objectives of environmental concern (including preservation of species, ecosystems and habitats), and that environmentalists should therefore focus on these relatively uncontroversial obligations, rather than, for example, on the intrinsic value of nature or on any distinctively biocentric duties. In view of the high value of future human survival and thriving, every generation is obliged to perpetuate such thriving, and the creative productivity of biotic systems which underpins human interactions with nature.[15] (Such a conclusion could also be derived from the findings of Chapter 4 of the present work about the significance of the human future.) A person choosing principles for intergenerational equity from behind a veil of ignorance with regard to the timing of his or her life, Norton suggests, would opt for constraints on 'trends that destabilize larger environmental contexts',[16] and thus endorse long-term obligations with regard to distant generations as well as to those of the immediate future. There need be no conflict between anthropocentrism and biocentrism, he claims, for their yield is the same. Norton calls the view that obligations concerning future generations and non-human nature tally in this way the 'convergence hypothesis'.[17]

Besides endorsing obligations to sustain human flourishing into the future (subject to the above provisos), we should recognise, with Norton, that obligations with regard to the further (non-immediate) future make a substantive difference. The arguments presented in Chapter 6 against conventional discounting support this stance; future costs and benefits should only be discounted where relevant grounds (such as uncertainty) exist, or obligations with regard to generations more than three decades away will be recognised in theory but disregarded in practice. These obligations were also central to the case for a global population policy, as argued in Chapter 7, and for preserving biodiversity, as argued in Chapter 8. Criticisms were supplied in Chapter 2 of arguments seeking to derive obligations through a Rawlsian veil-of-ignorance approach. Another objection to such contractarian thought-experiments concerns the assumption of interpretations making the contracting parties represent different generations, namely that the number of generations has somehow been determined beforehand, even though the principles adopted may actually play a large part in determining this number. A similar objection undermines Rawls' revised interpretation of his thought-experiment, presented in *Political Liberalism*, where he effectively assumes co-operation between generations independent of the making of the contract which establishes such co-operation.[18] Some parallel criticisms of contractarian derivations of these obligations have been presented by Robert E. Goodin.[19] But the obligations in question can readily be derived instead from the foreseeable consequences of present action for future lives and their quality, and thus the difference which most members of the present generation can make to the prospects of future generations, whether of humanity (as in Norton's argument) or of all living species.

However, the extent of the overlap between the yields of anthropocentric and biocentric ethics is manifestly far from total. There will be cases where no benefit to humanity from a species is either known or remotely likely, as with the preservation of many of the species of the bed of the deep ocean or living in the lake of fresh water two miles beneath the Antarctic ice-cap. Also there will be many cases where all the arguments from human interests are evenly poised, and arguments from non-human interests would accordingly prove decisive, unless disallowed; and there will be other cases where non-human interests, including those of wild creatures (such as interests in intactness of habitats or in freedom from suffering) are more considerable than any relevant human interests, and should be treated as such. This already shows that theories of environmental justice must extend beyond obligations with regard to future human beings, despite the deliverances of the convergence hypothesis.[20] Besides, if obligations relating to non-human interests sometimes have enough weight to outweigh obligations of justice relating to future humans, then they must count as obligations of justice themselves (albeit duties of imperfect obligation). But in any case, theories of environmental justice must also extend to the environmental needs of contemporary human beings, as noted above, and as the Environmental Justice Movement rightly insists; and these are needs which sometimes conflict with those of future generations. While the convergence hypothesis could in theory be extended to cover these contemporary human interests, this change would of itself require such a drastic restructuring that beginning with a quite different theory is preferable, such as the consequentialist and biocentric theory presented here.

Belief in humanity's stewardship or trusteeship can readily be combined with this ethical theory, which in turn supplies a substantive content to trusteeship (such as the policies argued for on a consequentialist and biocentric basis in Chapters 5 to 8).[21] Goodin, however, treats trusteeship theories as implicitly contractarian, and criticises them for embodying assumptions about reciprocity between present agents and posterity which disregard the facts that no one ever enters into such a contract, and that posterity, as remarked by Joseph Addison, seems never to do anything for us.[22]

While Addison's claim has been contested,[23] the possibility of posterity benefiting the people of the present (already discussed in Chapter 4) is not the central issue here. For while some religious versions of belief in trusteeship involve kinds of reciprocity between humanity and God, the view that current moral agents are trustees of nature does not depend either on their consenting to any form of contract or on any reciprocation on the part of our successors. True, the trusteeship view implies that succeeding generations will be trustees in their turn, but present obligations (as opposed to motivations) would be little affected, given this view, even if we were somehow to learn that future generations were unlikely to discharge their obligations as trustees. For adherents of the trusteeship view are free to hold, with consequentialists, that the grounds of the obligations of future people will correspond to the grounds of current obligations, consisting in the foreseeable consequences of their own actions, practices and policies; fortunately these grounds do not depend on the performance of their predecessors, any more than the grounds of our obligations depend on deeds of our successors, although their ability to discharge these obligations will in many ways depend on what we do and what we leave for them. Fortunately there is also no more reason to anticipate that future generations will fall short to a greater extent than the present generation.

Some more specific principles concerning relations between generations will be discussed in the next chapter. But equity between contemporaries needs to be considered first, both for its own sake and because provision for it could make a vital difference to the future.

JUSTICE WITHIN GENERATIONS

Current needs are usually discoverable with greater certainty than future needs, and are often more amenable to satisfaction. This means that a greater difference to the satisfaction of needs can often be made by addressing the needs of the present, environmental needs (such as the need for a decent environment) included. Present needs, however, should not be prioritised ahead of future needs as such; for action relating to the needs of coming decades (in matters of supplies of fresh water, control of global warming and avoidance of over-population, for example) can often make a greater difference than short-term measures in these regards, and action (say) to preserve species or to curtail nuclear pollution could easily prove vital for hundreds or even thousands of generations. Basic needs of the future dependent on current action for their satisfaction should receive priority ahead of lesser needs of the present, at least on a par with current basic needs; indeed delivering basic needs, wherever they are located, is a basic obligation of justice, perhaps because it makes the greatest available difference in terms of values and disvalues, goods and evils. So, at least, consequentialists contend. (This is not the place to analyse the concepts of needs, basic and otherwise, particularly as I have undertaken this task elsewhere.)[24]

Yet it remains important that ours is the last generation which can deliver current needs, whereas the delivery of some future needs (for example, through technological research) can be shared cumulatively with our successors. Further, the rectification of current injustices is often a prerequisite for environmental justice in future generations. Thus the processes involving transfers of pollution and of unhealthy work generated by rich and powerful corporations to poor neighbourhoods and poor countries are likely to be exacerbated in future unless they are redressed in the present, and unless institutional changes are adopted to prevent their recurrence, and to bequeath more equitable social and international relations to posterity. Again, generations which inherit large inequalities of wealth between landowners and peasants (current in much of Latin America) are likely to suffer in intensified form the environmental problems already associated in the present with oppression and poverty in many Third World countries.

Indeed, attaining equity between and among current peoples and among contemporaries in general, with all that this involves, is likely to be necessary for the introduction of equity between successive generations, because only if equitable arrangements are established and transmitted are there likely to be equitable societies, locally or globally, in decades to come. This could mean the enhancement of urban environments, particularly in poor districts, in the present; urban environments are now the environments of over half of humanity. It could also mean massive assistance to the Third World to allow poor countries to cope with their problems of underdevelopment and of environmental degradation; for without such assistance, there is little prospect of same-generation or intragenerational equity in those countries, let alone of between-generations or intergenerational equity.[25]

Before most Third World countries can invest significant resources in tackling environmental problems, their burden of debt needs to be lifted. The scale of the problem can be seen from the fact that during the 1982–90 period, Southern countries paid creditor countries $1345 billion by way of debt servicing, considerably more than total resource flows to the South.[26] Their interest payments alone through that period came to a monthly average of $6.5 billion.[27] Many of these debts can never be repaid; Mozambique, for example, owed nearly five times its gross national product.[28] Meanwhile debt repayments divert resources from delivering basic needs; Uganda, for example, spends $2.50 per person on health care and $15 per person on debt repayments.[29]

Unsurprisingly, there is a high correlation between indebtedness and environmental degradation such as destruction of forests, for deforestation is often a resort of afforested countries in need of revenue,[30] albeit abetted by transnational corporations and international banks. Thus Susan George points out that all sixteen of the most heavily indebted countries with forests are among the big deforesters; and that five of the eight most heavily indebted countries (Brazil, India, Indonesia, Mexico and Nigeria) figure in the top ten in the deforestation league.[31] Also rates of deforestation have accelerated through much the same period as the intensification of the debt crisis.[32] Thus indebtedness almost certainly contributes to environmental deforestation, and probably to other forms of environmental degradation too, such as desertification and urban pollution, as short cuts are taken with nature to service debts.[33] Further, Structural Adjustment Programmes, imposed by the International Monetary Fund on indebted countries (including twelve which became major deforesters), involve strict conditions including increasing exports and reducing domestic spending, and thus tend to distort development and to exacerbate social and environmental problems.[34] Meanwhile all countries suffer from the global warming and biodiversity loss consequent on deforestation (see Chapters 5 and 8).

While a few Third World countries may be prosperous enough to service their debts, the majority cannot be expected to do so. These debts were often incurred by unrepresentative, long-departed, often corrupt regimes, at times when bankers were offering surplus revenues on deceptively attractive terms. In any case their impacts seriously blight the life-chances of billions of people, many born since the debts were undertaken. To insist on repayment in such circumstances is inequitable. Current international initiatives to relieve a few of the most heavily indebted countries of their debts are proving excessively slow moving and inadequate in extent; nothing less is needed than a complete once-off debt forgiveness for all countries unable to repay, rather as advocated by the pressure-group Jubilee 2000, but with conditions in some cases concerning human rights and environmental protection and conservation. (A once-off debt cancellation in, say, 2000 for countries unable to repay their debts would be consistent with loans continuing to be made, where appropriate, to Third World and transitional countries capable of repaying them.) While international debt relief alone will not solve developmental or environmental problems, and a restructuring of international relations and policies is also needed, these problems cannot be solved without it;[35] and this is a measure likely to bring considerable benefits both to present and future generations, and both to humanity and other species.

Nor is debt all owed by the South to the North. As mentioned in Chapter 7, past Northern exploitation of Southern countries, through colonialism and the subsequent North-dominated trading system, together with the considerable Northern role in generating ecological damage in the forms of climate change and biodiversity loss, has led to claims that the North owes an 'ecological debt' to developing countries; hence the North should restructure its trading patterns, transferring much-needed technology and funding biodiversity preservation as well as cancelling unpayable debts (in the more conventional sense).[36] Such claims partly involve appeals for compensation, and partly for more equitable treatment in the present with a view to future needs and opportunities. While consequentialism is centrally concerned with outcomes such as these, it can also justify patterns or systems of compensation and reparation, which can rectify relationships and discourage future exploitation. In the current context, duties of compensation are undeniable, despite the refusal of Northern delegations to recognise this in the Rio Biodiversity Convention. But in any case current problems (including ecological problems) and the current and foreseeable needs of Southern countries support the policies just mentioned as a matter of equity, and also support priority being given within these policies to the countries with the gravest problems. In many areas (such as biodiversity preservation) there are strong realist arguments from Northern self-interest for such measures as well.

This section obviously leaves much unsaid about global justice within generations, even in environmental contexts. Nevertheless it at least serves to correct the views that environmental justice predominantly concerns fairness to the future, or that it largely concerns nature preservation as opposed to protection and enhancement of the environments of contemporary people. If environmental justice means anything, it concerns the environments of the poor, of women, and of people of Third World countries, as the Environmental Justice Movement affirms;[37] as I have argued, the related obligations sometimes take precedence over those focused on the future or on nature preservation.

JUSTICE AND THE ENVIRONMENT AS PROPERTY

The property system held a central place in David Hume's theory of justice,[38] because of what he took to be its overall utility. In the twentieth century, Terry L. Anderson and Donald R. Leal have advanced the view that property rights need to be extended right across the natural environment, as the property system is uniquely efficient in ensuring care for what is owned, although they reject the goal of sustainable development.[39] While their arguments seem capable of supporting ownership in many situations on the part of a local community as much as private ownership, private ownership is undoubtedly what they have in mind.[40] Some of the related issues (such as that of Intellectual Property Rights) have already been introduced in Chapter 8; however, some brief comments on the relation of property to the care of the environment are in place in a chapter on justice and world order.

Efficient as the property system often is, enabling environmental costs to be taken into account much more reliably than open-access systems would be likely to do, and sometimes fostering the kind of pride which ownership can bring, a strong case can also be made (based, for example, on the needs of international shipping) for much of the biosphere remaining a commons, belonging neither to states as sovereign territory nor to individuals or corporations as property.

Strong arguments can also be deployed for large tracts of land to remain in communal rather than private ownership (and not only land). Land, in particular, is in such short supply in many countries that allocating all unowned land to proprietors and subjecting it to laws of sale, gift and inheritance would include among predictable outcomes depriving many people of all legitimate access to land, with foreseeable adverse consequences both for society and for the land as well. However, where cultivable land is already held in large and often unproductive estates, as in much of Latin America, there is a strong case for redistributing it to landless people in lots large enough to support a family. Small proprietors have been found to care for the land better than large estates, and also to make a better contribution to food production.[41]

Currently most of the atmosphere, the stratosphere, the seas and the oceans (other than off-shore Exclusive Economic Zones) are neither anyone's property nor any country's territory, while international treaties restrict the uses of Antarctica even for the countries which claim parts of it. Shared access to these global commons, tempered by regulatory international agreements, almost certainly forms a better system than one of division into territorial or proprietorial zones; such agreements are needed, for example, to curtail pollution and to regulate shipping. This system, however, could be enhanced, as suggested by Christopher D. Stone, through the appointment and international recognition of guardians of the oceans, endowed with the standing necessary to initiate legal and diplomatic action on behalf of ocean biomes or ecosystems at risk of damage from pollution or overfishing, in cases where, if the ocean were a state, such action could have been taken with reasonable prospects of redress or relief.[42] (The international bodies instituted under the Law of the Sea Treaty from 1994 could receive these powers.) Stone has also proposed a tax on all uses of the global commons, the proceeds to be used for global conservation and repair.[43] Unfortunately this admirable suggestion conflicts with the proposal of Michael Grubb, adopted above in Chapter 5, to distribute the absorptive capacities of the atmosphere to states in proportion to their population, and allow trading of part of these quotas; and Grubb's proposal could well be needed to bring developing countries within the Kyoto regime. Stone's tax could still be implemented for other uses of the resources of the commons, such as fishing and waste-dumping in the oceans,[44] and stationing satellites in space above the Earth's equator.[45] Indeed, as was argued in Chapter 5, where resource-use amounts to resource-depletion, this should be done.

Open and unrestricted access to lands, as opposed to communal ownership, can certainly lead to environmental degradation (and no more than this can safely be concluded from Garrett Hardin's celebrated essay 'The Tragedy of the Commons').[46] But so can private ownership, a fact which adds to the importance of Goodin's argument that none of the justifications of property rights confer on owners a right to destroy, and that where assets are irreplaceable, owners have an obligation to preserve them.[47] Besides, communal ownership is the traditional basis of cultivation and of grazing in much of the Third World. This system often fosters communal pride in the shared land, and concern to prevent its degradation; continuation of this system is also crucial for the preservation of cultures, and thus of cultural diversity. The enclosure of communal land is unlikely to foster social justice, or to preserve biodiversity, any more than it did in Britain in the eighteenth century. The same applies to attempts to establish national parks on lands which were previously communal and then to deny local people access to them.[48]

Communal ownership could be recognised in knowledge of plants and wildlife and related skills, as Markku Oksanen has suggested,[49] as well as in land, subject to limits restricting such ownership to distinctive current knowledge (perhaps subject to time-limits), as opposed to knowledge long since exported. While patents might be inappropriate for knowledge which has developed across many generations, other forms of intellectual property rights could be devised for such communal knowledge, and its recognition could assist both local communities and their preservation of local biodiversity. The apparently attractive view that unowned natural resources are the common heritage of humanity, besides its anthropocentrism, carries the liability that in the absence of institutional safeguards it is prone to hand biological resources direct to corporations which are then free to patent them; hence the ideal of global sharing becomes a cloak for 'biopiracy', the undermining of local cultures (not all of which, admittedly, should be preserved) and sometimes the disappearance of biodiversity.[50]

The current world situation about biological patents is becoming extremely complex, particularly where patents for plants are concerned. Plant breeders do not strictly require patents, but often behave as if patents are indispensable;[51] views about the merits and demerits of patents and of alternative intellectual property regimes are diverse in the extreme;[52] and while not all countries recognise the GATT agreement on Trade Related Intellectual Property, those which do not recognise it are increasingly being put under pressure to do so.[53] The Ottawa-based Crucible Group (an international think-tank) recommends that the United Nations convene an international conference on society and innovation to discuss the ethics and international regulation of such matters;[54] in the circumstances such a multilateral conference has much to commend it.

WORLD ORDER AND NATIONAL SOVEREIGNTY

Like global ethics, environmental problems are no respecters of national boundaries. From sharing the waters of the Euphrates, the Nile and the Colorado to global warming, they require international co-operation, and often international funding. International bodies, however, seldom have sufficient authority or resources, or sufficient independence from the national powers which form them, and the outcomes are often avoidable harm to human beings and other creatures of the present and the future, and to the ecosystems on which they depend. Regarded as maldistribution, such outcomes can comprise environmental injustice. The issues thus arise of whether a system of sovereign states is equal to these problems, and of whether cosmopolitans can any longer support it.

The role of cosmopolitans in a world with far from ethical structures will be discussed in the final chapter, but the ethics of political sovereignty must be introduced here because of the relevance of this world system to global environmental justice. World problems are not confined to the environmental problems already discussed, or to the economic problems which often underpin them. The recurrent problems of territorial disputes and armed conflicts (civil wars included) generate or exacerbate vast environmental problems, and displace millions of their survivors, quite apart from the carnage that they visit on humanity. Not even the Nuclear Non-Proliferation Treaty has been able to prevent the spread of nuclear weapons or to terminate nuclear tests. Since most armed disputes are either disputes between states or struggles for the control of a particular state, they strengthen the case for consideration of an alternative system, that of world government. L. Jonathan Cohen has presented working towards this system as an obligation.[55]

Yet world government carries the risk of even worse eventualities. The world sovereign would rule a world empire, immune from external challenge; and might do so in a manner liable to foster continental rebellions or global coups; while civil wars within this system would be world wars. A democratic global system could well prove a great gain, but would be at risk of overthrow in favour of a global tyranny, and of subversion by global corruption. Changes of government, whether at regional or global level, might be prohibitively difficult to secure, even when seriously overdue; and the best-devised global system of checks and balances might fail, with devastating planet-wide effects. Clearly other systems should be considered.

A range of systems could be devised in which international co-operation at both regional and global levels would increase, in which global environmental problems would be internationally regulated and policed, and in which the necessary powers and resources were delegated, subject to safeguards, to international agencies. Such systems would be compatible with the continued existence of national states, except that the nature of sovereignty would gradually change, being pooled for regional or sectoral purposes in a range of treaties,[56] some of which would be irrevocable. People's sense of national identity (and the sense of cultural identity which often accompanies it) could be preserved, but the significance of this too would be modified as people increasingly came to belong to international or global societies or networks,[57] and to take on additional or new forms of identity. In the course of time, the monopoly of force, currently vested almost entirely in sovereign states, would also be modified, so as to allow of international peace-keeping and peace-making forces, to which states would contribute in proportion to their resources. Thomas Pogge has offered constructive suggestions for moving from where we now are to such a multilayered system of global governance or of 'institutional cosmopolitanism'.[58]

Far from being fanciful, the kinds of changes depicted in the previous paragraph are actually beginning to happen. They have their dangers, against which precautions have to be taken, but in many cases the dangers are less considerable both in magnitude and probability than the recurrent disasters and the awesome catastrophes of the twentieth century. An example of current processes with the potential to develop much further in these directions is supplied by the Kyoto agreement of 1997 and its aftermath.

An agreement to control and regulate greenhouse gas emissions is indispensable in the modern world, even one such as that made at Kyoto which reduces total emissions by less than 6 per cent from 1990 levels, which bases authorised national emissions on historical (1990) amounts (thus rewarding big historical polluters), and which permits wealthy would-be polluters to purchase emission quotas from states with surplus entitlements. But the agreement will be ineffective unless a powerful regulatory body is established, empowered to police compliance as well as to authorise emissions trading. Powers which were not among the Kyoto signatories need to become involved in the ongoing negotiations, even though many Third World countries should be granted quotas for increased emissions so as to provide for the basic needs of their citizens,[59] and limits to trading need to be agreed to prevent the alienation of this element in national quotas. In amendments to the agreement, quotas should be proportioned not to historical emissions but to population, as has long been proposed by Michael Grubb,[60] since each person alive has as strong a moral entitlement to benefit from global absorptive capacities (or sinks) as any other. At the same time, the total of permissible emissions should be significantly reduced, for the sake of the inhabitants of small islands like the Maldives and of coastal plains and estuaries like those of Bangladesh. If acceptable treaties can be agreed, they should in due course be made permanent, and the power to enforce them transferred to a representative international authority.

In future, humanity and fellow species will be dependent on such international treaties and regimes. Recognition of human stewardship or trusteeship of the planet involves a readiness to play a full part, and this applies to all countries. Playing a full part does not involve uniform responsibilities, but responsibilities differentiated by the differing powers and resources of the various states or regional groupings, as the discussion of sustainability in Chapter 10 underlines. Nevertheless, global trusteeship supports this kind of pooling of sovereignty, and without it, and without the consistent application of such agreements at all political levels, national and local included, environmental justice is unlikely to be done.

NOTES

1. United Nations Conference on Environment and Development, 'Rio Declaration on Environment and Development', in Wesley Granberg-Michaelson (ed.), *Redeeming the Creation*, 86–90, Principle 1, p. 86.
2. Ibid., Principle 3.
3. Brian Barry, *Liberty and Justice: Essays in Political Theory 2*, p. 259.
4. See Carol Gilligan, *In a Different Voice: Psychological Theory and Women's Development*.
5. Carl Talbot, 'Environmental Justice', *Encyclopedia of Applied Ethics*, vol. 2, 93–103.
6. Robin Attfield, *Environmental Philosophy: Principles and Prospects*, ch. 16.
7. Peter S. Wenz, *Environmental Justice*.
8. Ibid., p. 333.

9. Ibid.
10. Ibid., p. 328.
11. Robin Attfield, *The Ethics of Environmental Concern*, ch. 9; Donald VanDeVeer, 'Interspecific Justice', *Inquiry*, 22, 1979, 55–79.
12. Attfield, *Environmental Concern*, pp. 172–7.
13. Farhana Yamin, 'Biodiversity, Ethics and International Law', *International Affairs*, 71.3, 1995, 529–46.
14. Paul Taylor, *Respect for Nature: A Theory of Environmental Ethics.*
15. Bryan G. Norton, *Toward Unity Among Environmentalists*, p. 216.
16. Ibid., p. 217.
17. Ibid., pp. 240–3.
18. John Rawls, *Political Liberalism*, p. 274.
19. Robert E. Goodin, *Protecting the Vulnerable: A Reanalysis of Our Social Responsibilities*, pp. 170–4.
20. See also Robin Attfield, 'Development and Environmentalism', in Robin Attfield and Barry Wilkins (eds), *International Justice and the Third World*, 151–68.
21. See further Robin Attfield, 'Environmental Ethics and Intergenerational Equity', *Inquiry*, 41.2, 1998, 207–22.
22. Goodin, *Protecting the Vulnerable*, pp. 174–7; Joseph Addison, *The Spectator*, no. 583, 20 August 1714.
23. John O'Neill, *Ecology, Policy and Politics: Human Well-Being and the Natural World*, pp. 26–36.
24. Robin Attfield, *A Theory of Value and Obligation*, chs 4, 5 and 8; *Value, Obligation and Meta-Ethics*, chs 5, 6 and 9.
25. See further the parallel passages in Attfield, 'Environmental Ethics and Intergenerational Equity'.
26. Susan George, *The Debt Boomerang: How Third World Debt Harms Us All*, p. xv.
27. Ibid., p. xiv.
28. Jubilee 2000, *Fact Sheet on International Debt*, p. 2.
29. Ibid., p. 1.
30. George, *The Debt Boomerang*, ch. 1.
31. Ibid., p. 10.
32. Ibid., p. 11.
33. Ibid., p. 1.
34. Ibid., pp. 2 and 14–16.
35. Ibid., pp. 28–30.
36. Parvez Hassan, 'Moving Towards a Just International Environmental Law', in Simone Bilderbeek (ed.), *Biodiversity and International Law*, p. 75; Fidel Castro, *Tomorrow is Too Late*, pp. 19–24.
37. See Talbot, 'Environmental Justice'.
38. David Hume, *A Treatise of Human Nature*, ed. L. A. Selby-Bigge, bk III, pt 2 ('Of Justice and Injustice').
39. Terry L. Anderson and Donald R. Leal, *Free Market Environmentalism.*

40. Markku Oksanen, 'Privatising Genetic Resources: Biodiversity Preservation and Intellectual Property Rights', unpublished paper presented to European Consortium for Political Research Joint Sessions, University of Warwick, 1998.
41. Lester Brown and Erik Eckholm, 'Food Supplies', in Elizabeth Stamp (ed.), *Growing Out of Poverty*, 20–33, p. 28.
42. Christopher D. Stone, *The Gnat is Older than Man: Global Environment and Human Agenda*, pp. 83–8.
43. Ibid., pp. 208–20.
44. Ibid., p. 209.
45. Ibid., pp. 210–11.
46. Garrett Hardin, 'The Tragedy of the Commons', in John Barr (ed.), *The Environmental Handbook: Action Guide for the UK*, 47–65.
47. Robert E. Goodin, 'Property Rights and Preservationist Duties', *Inquiry*, 33, 1990, 401–32.
48. Smitu Kothari and Pramod Parajuli, 'No Nature Without Social Justice: A Plea for Cultural and Ecological Pluralism in India', in Wolfgang Sachs (ed.), *Global Ecology*, 224–41; see also Ramachandra Guha, 'Radical American Environmentalism and Wilderness Preservation: A Third World Critique', *Environmental Ethics*, 11, 1989, 71–83.
49. Oksanen, 'Privatising Genetic Resources'.
50. Vandana Shiva, *Biopiracy: The Plunder of Nature and Knowledge*, pp. 65–85, 96–9.
51. The Crucible Group, *People, Plants and Patents: The Impact of Intellectual Property on Biodiversity, Conservation, Trade and Rural Society*, p. xx.
52. Ibid., pp. 53–94.
53. Ibid., pp. xix-xx; 83–4.
54. Ibid., p. xvi.
55. L. Jonathan Cohen, *The Principles of World Citizenship.*
56. See Commission on Global Governance, *Our Global Neighbourhood*, which endorses such a qualified abridgement of national sovereignty.
57. As depicted in Janna Thompson, *Justice and World Order: A Philosophical Inquiry*, ch. 9.
58. Thomas Pogge, 'Cosmopolitanism and Sovereignty', *Ethics*, 103, 1992, 48–75.
59. Henry Shue, 'Equity in an International Agreement on Climate Change' (unpublished), paper presented to IPCC workshop on 'Equity and Social Considerations Related to Climate Change', Nairobi, 1994, pp. 7–14.
60. Michael Grubb, *The Greenhouse Effect: Negotiating Targets*; *Energy Policies and the Greenhouse Effect.*

CHAPTER 10

SUSTAINABILITY: PERSPECTIVES AND PRINCIPLES

INTRODUCTION

It might seem that a chapter on moving towards sustainability should seek to derive the best objective policies from science and from principles of sustainability and equity, with a view to the best overall outcomes emerging from their worldwide adoption. A consequentialist and cosmopolitan ethic might appear to demand nothing less, perhaps advocating in addition principled early interventions to prevent irreversible hazards without waiting for scientific confirmation. Yet such a top-down approach would also be likely, even if such policies were to be adopted, to generate new problems of dissent and non-compliance, with accusations of the imposition of dominant ideology dressed up as universal values. While this is a hazard in each political community, it is all the more a problem if applied to international society, where, in the absence of any central authority, policies and actions must perforce turn on multilateral agreement, not only achieved in well-publicised international conferences and treaties but also sustained throughout the period of the delivery of a given policy, and must depend on the good will of all the states concerned. Such an approach also accords too little recognition to the importance of the autonomy both of individuals and of countries, and of its translation into freely chosen outcomes (a recognition which consequentialists can consistently make and would widely wish to see taken seriously).

Nothing less than an equitable worldwide system of decision-making could solve the problem identified here; and even that might prove insufficient, in view of the imbalances of power and of access to information which could in theory persist and coexist with such a system. But the depiction of such a system is in any case well beyond the scope of this book; and fortunately the non-existence of such a system need not impede the more modest task of outlining principles and policy directions which ought to be adopted with a view to tackling world problems, however inadequate may be the international framework in which they are likely to be tackled for the foreseeable future. Nevertheless some recognition can be shown here of the problems confronting attempts at global planning and global policy-making. This will involve reflection on the diversities in perceptions of global environmental problems and of what makes these problems global, and accordingly on the complexity of the second-order problem of seeking agreed solutions to problems which are perceived differently (the theme of the next two sections). But much else will be needed before global planning can have the necessary quality and deserve general confidence; one of many desirable changes is likely to be a spread of participatory democracy, which is going to be needed to generate informed and freely chosen decisions and policies at all the relevant levels of decision-making (see Chapter 11).

PERCEPTIONS OF GLOBAL PROBLEMS: SURVIVAL AND GLOBAL WARMING

Diverse perceptions of global environmental problems have been implicit in the above chapters on subjects ranging from resources to biodiversity. Even the issue of securing human survival generates rival perspectives, some focusing on the survival of the current generation in parts of the Third World, faced (for example) with serious and growing shortages of fresh water, others on the impending threat for the coming decades of inundation of islands and low-lying areas, generated by global warming, and yet others on survival in future centuries and the resources needed to foresee and forestall catastrophes such as the onset of new pandemic diseases or collisions with asteroids. These issues are all deserving of serious attention, without standing outside the bounds of rational comparison and appraisal; yet it remains important that the problem of survival generates such diverse interpretations, all capable of being taken seriously.

Differing attitudes to resources are often attributable to differing perspectives from North and South. Take carbon emissions. As Steven Yearley remarks, the problem can be presented and perceived from a Northern perspective as due to big polluters, with Brazil, China and India included among the top six (after USA and the former USSR). But in terms of carbon emissions per head, those of China and India are still modest, and those of Brazil lower than those of Germany, Canada, Japan, United Kingdom and France.[1] This has been pointed out by Anil Agarwal and Sunita Narain of the Centre for Science and Environment, New Delhi, who also maintain that the problem cannot be seen to consist, without distinction, in either carbon or methane emissions regardless of source, or to be captured by the simple aggregation of, for example, emissions from powerful cars used for inessential journeys and emissions from rice fields cultivated for subsistence farming. A distinction is necessary, they suggest, between the 'survival' emissions of the poor and the 'luxury' emissions of the rich.[2] The distinction could be better articulated, for example as a distinction between pollution relating to basic needs and pollution not so related, or between unavoidable and optional greenhouse pollution.[3] But some such distinction is important if the poor are not to be blamed for doing what subsistence demands, in cases where this leaves no room for choice.

The Kyoto agreement could be seen as taking Agarwal and Narain's first distinction (between emission totals and rates) into account, in that countries with higher per capita emission rates are expected to accept bigger cuts from the levels of their emissions of 1990. But its full recognition would involve moving to a regime of quotas proportioned to population (or some modification, devised to avoid rewarding population growth), as argued in Chapter 5. Moves in the direction of such a regime may well become necessary in order to bring the South into the Kyoto regime. This objective should also involve recognition of Agarwal and Narain's second distinction; national quotas, as suggested in that chapter, should include inalienable quotas for emissions relating to basic needs. Thus the perspective introduced by Agarwal and Narain (and much debated in the intervening years) proves important both in identifying where the problem of global warming lies, and in attempts to resolve it.

The debate also serves to disclose, as Yearley points out, that the distinction between anthropogenic and other emissions is less than clear. For example, when anthropogenic acid rain reduces biomass and thus carbon sinks, does this or does it not count as a corresponding increase of the anthropogenic carbon emissions from the country causing the acid rain? This distinction in effect concerns moral responsibility and transcends the scope of science; it involves making ethical judgements[4] such as the increasingly recognised principle that the polluter should pay. Not just anything counts as a reasonable ethical judgement (try comparing this principle with its negation), and I have no hesitation (unlike Yearley, perhaps) in holding that ethical judgements can aspire to objectivity. But in the context of international agreements, such issues can only be resolved by political negotiation about ethics and, relatedly, about the common interest.

PERCEPTIONS OF GLOBAL PROBLEMS: OZONE

Global warming is far from the only case of diverse perspectives about the nature of a global problem, or about what makes such a problem global, and about the need for negotiation between perspectives. This will be borne out here by a more detailed consideration of the issues raised by CFCs, already discussed in Chapter 5 in connection with the need for international co-operation. In this matter, the Toronto group, including USA, Canada, Norway and Sweden, located the problem in non-essential use (spray cans, for example), and wanted an international ban on such use, while the European Union argued for a cap on overall production instead, regarding the amount (as opposed to the fact) of CFC production as the problem. The Montreal Protocol itself (1987) required a freeze by 1990 of overall production at 1986 levels, with cuts thereafter, but allowed developing countries a ten-year grace period before introducing cuts, and a modest increase of consumption during that period to meet 'basic domestic needs'. Trade in CFCs with non-signatories was effectively banned by the Protocol as from the early 1990s. The Protocol also provided for its own review in case new evidence indicated that its terms were too weak (or too strong); and new evidence, published in 1988, led to the London Revisions, which required all CFCs (as opposed to substitutes) to be phased out by 2000.[5]

While the evidence (of ozone holes in both Antarctic and Arctic) substantially undermined the earlier perspectives both of the Toronto group and of the European Union, debate continued within USA about whether the solution lay in regulation or in self-help (such as the use of sun-block). But environmentalists were able to point out that wildlife could not be protected by such measures, and that many humans were exposed to the sun not through choice but because of the nature of their employment. Hence the libertarian position was convincingly worsted, and the US government supported the agreements of 1987 and 1990 which gave most developing countries little option but to sign up to a ban, granted that non-signatories were subject to sanctions.[6] But the Protocol left India and China free to manufacture for domestic use, and this they were poised to do. Indeed, if they had proceeded to use their Protocol quotas to the full, they could have ruined the entire project of emissions reduction. But forgoing the manufacture of CFCs meant for them losing their investment unless compensated and offered alternative technology. So a compromise had to be reached in the London conference, involving fair technology transfer and large-scale financial aid to assist developing country compliance.[7]

The solution reached was thus a negotiated compromise. Without endorsing Yearley's view that the agreement was not informed by agreed, objective science,[8] we should recognise that science did not (and could not) determine which solution was in the global interest.[9] National interests, as perceived in the different countries, obviously diverged, as in some cases did perceptions of the national interest of USA. Perceptions of the global interest diverged as well in the light of different values, all these views being adopted in full awareness of the developing body of scientific knowledge. Many principles were at stake, principles of equity, efficiency, compensatory justice and provision for basic needs; indeed the eventual explicit provision for the latter forms an encouraging precedent. The range of relevant principles (all recognisable as such on the part of cosmopolitan ethicists) is enough to account for differences of perspective, quite apart from divergences of self-interest. Through the negotiating strength of India and China, developing countries fared better than they might have done. Northern countries veered from willingness to abstain from virtually any regulation to willingness to pressurise developing countries and each other to accept a strong international regulatory regime. Nevertheless the outcome was a combined triumph for science, environmental NGOs and diplomacy, and displays the possibility of finding sustainable solutions, divergent perspectives notwithstanding.

SOME PRINCIPLES OF EQUITY AND SUSTAINABILITY

It is salutary to return to the subject of principles of equity and of sustainability in the light of the above histories of recent international negotiations, and also in the light of the potential role of ethical principles as well as of science and of the realities of practical politics in moulding policies. Theorists have advanced a range of principles of intersocietal and of intergenerational equity, and now is the time to discuss such principles (and also to discharge the promise made in Chapter 5 to give further consideration to obligations relating to resource needs).

Working within a contractarian framework, Charles R. Beitz has suggested that parties subject to a veil of ignorance and unaware of 'the resource endowments of their own societies . . . would agree on a resource redistribution principle that would give each society a fair chance to develop just political institutions and an economy capable of satisfying its members' basic needs'.[10] My reservations about reliance on a contractarian approach were explained in Chapter 2; however, working with this approach sometimes carries a strong heuristic value, particularly where, as here, the distinctive environment-related weaknesses of this approach (concerning future generations and non-human species) are not in evidence.

Indeed, if enough qualifications are made, this principle of inter-societal equity has much to commend it, and could be endorsed by consequentialists who prioritise the satisfaction of basic needs. One qualification concerns the need to redistribute much more than resources to facilitate just political institutions and sustainable economies. Reservations also concern the meaning of 'redistribution of resources', since most resources (forests, fisheries, fresh water) cannot realistically be redistributed. Such redistribution must concern recurrent flows of financial resources to facilitate good government and sustainable economies, despite the unevenness and arbitrariness with which natural resources are distributed in the world as we find it. A redistribution of resources alone could still prove insufficient, but in combination with other measures such redistribution could be crucial (see Chapters 6 and 7). It might involve measures already commended above, such as debt relief and aid to facilitate biodiversity conservation, together with other concessionary aid from North to South, and more favourable terms for the trade of Southern countries.

Sustainability in the South, as Michael Jacobs has argued, involves increased and more stable prices for the primary commodities on which many Third World countries depend, and of the manufactured goods which they export to the North. For manufactured goods this involves a reduction in Northern protectionism. Ideally it would also involve assistance for economic diversification and moves away from reliance on a limited range of primary commodities and in the direction of more commodity processing.[11] Such aid would be less generous than it sounds, and not only because donors of official aid usually benefit through increases in trade. For acceptance of the 'Polluter Pays' principle (a principle which should be accepted, I suggest, except where polluting is unavoidable or where responsibility is mitigated, or where the beneficiaries of pollution are lacking in resources) would imply that those who have benefited from the degradation of environments in the South, and the consequent social problems (namely the countries, financial institutions and multinational corporations of the North), should bear the costs of rectifying the consequent damage, and that these costs should be regarded as a form of compensation.[12] The willingness of Northern countries to enter into international agreements about CFCs and global warming attests a partial recognition of this principle, in sectors where Northern responsibility is hard to contest; for the argument from compensation often secures recognition even where arguments from global justice go unheeded. In any case, the same principle applies to restoring environmental degradation, and (if generalised to include rectifying the harm the North has caused) to alleviating the dependence on single primary products into which many Southern countries have been thrust, and thus to putting in place some of the key missing economic conditions of sustainable development.

As Jacobs suggests, greater environmental controls on transnational companies operating in the South are another prerequisite for sustainable development there. Southern countries are often in a weak bargaining position in dealings with these powerful companies, and the introduction of environmental standards would involve regulations or codes of conduct at the level of international treaties or at least at the level of the European Union.[13] The Multilateral Agreement on Investment, temporarily abandoned by its proponents but all too likely to be resuscitated in a new framework, would have the opposite tendency of strengthening the position of foreign companies worldwide, and should be replaced by a treaty strengthening their responsibilities, environmental responsibilities included. Jacobs further mentions the need for reappraisal of the environmental bearing of the World Bank and its projects;[14] without doubt the same applies to the environmental profile of the International Monetary Fund and of its financial packages, which have often undermined systems of health, welfare and education, and fostered large infrastructural projects such as giant dams, with lamentable environment consequences. Jacobs is also correct in urging a diversion of government budgets away from military spending (a change which the IMF could usefully encourage). Much of the world military spending (of $900 billion in 1991) reduces rather than enhances security; if a fraction of it were diverted to measures of 'environmental security' (population planning, water management, reforestation and the like),[15] the prospects for sustainable development would in many countries be hugely enhanced.

Meanwhile Michael D. Young, who puts forward some valuable principles of intergenerational equity, sensibly maintains that an important condition of such equity would be 'an efficient, diversified and ecologically sustainable economy'.[16] Certainly an ecologically sustainable economy is likely to be a necessary condition of intergenerational equity, and of interspecies justice too, at regional, national and global levels alike, far removed as this state of affairs is from most actual economies; while economies efficient enough to supply the material needs of their society are likewise a prerequisite of social justice. The issue of the shape of such a global economic regime lies beyond the scope of this book. Yet international agreements on limiting emissions of CFCs and generally of greenhouse gases comprise important steps towards such a regime, as would systematic introduction of energy generation from non-fossil and non-nuclear sources, and enhancement of the terms of production and trade for Southern countries; so at least some of the prerequisites of this crucial state of affairs have been considered here. But since there is a potential clash between Young's criteria of efficiency and sustainability, it is appropriate to stress that the criterion of efficiency must be understood as constrained by that of sustainability, if sustainable economies are eventually to be attained. Some of the other principles discussed or introduced by Young will be considered in the coming section.

SOME PRINCIPLES OF INTERGENERATIONAL EQUITY

One principle considered by Young (who digs deeper here than *Our Common Future*, discussed in Chapter 6) is the Pareto principle which requires that changes should make no one worse off than they would have been otherwise. Expressed like this, however, the principle would usually block all redistribution of resources, desirable as well as undesirable. Besides, if this principle is re-expressed in terms of the mere possibility of compensating the victims of change, it falls short of equity unless actual (and not just possible) compensation would be made. However, a revised form of the principle could justifiably be applied to cases where the beneficiaries of a change are in the present and the foreseeable victims (and the environmental costs) are spread out across future generations. While such victims could sometimes in theory be compensated for being subjected to risks of (say) low-level radioactivity from the residues of a nuclear power station, perhaps by some very long-term investment devised for this purpose, the prospects that this would happen for thousands of generations are minimal, and accordingly the change (in this case, the installation of a nuclear energy plant) should not be made, unless the prospects for safe decommissioning of nuclear plants and for safe burial of nuclear wastes undergo a dramatic transformation. In an intergenerational context, a defensible form of the principle would require that serious risks of harm to future generations (where people of this generation, whatever their identity, might have lived immune from this harm) must be likely to be compensated, and that avoidable changes should not be made where this requirement is not satisfied. (Young's interpretation of the Pareto principle, requiring that future generations actually will be compensated,[17] is however too strong, as decision-makers could never know that it would definitely be satisfied.) But none of this says anything about cases where both action and inaction carry long-term risks, as when the only choice is between new experimental technology and old hazardous technology.

As was recognised in Chapter 5, the depletion of scarce resources carries an ethical requirement of compensation to deprived parties, including affected future generations; I went on to suggest that this involves not, as suggested by Brian Barry, provision for matching opportunities in each generation, but an obligation in the present to provide, where possible and subject to certain provisos, for the satisfaction of foreseeable basic needs in the future. Others have suggested that resource-depletion must be compensated by increases in technological development or capital investment;[18] but polluting technology would be no compensation, particularly as future generations cannot make the authors of this pollution pay for it. Where the depletion to be compensated is of fossil fuels, compensatory technology would have to facilitate replacement supplies of energy without generating significant pollution; hence the obligation to compensate turns out to support research on and introduction of renewable (but non-nuclear) energy generation. Some resource-depletion, however, such as losses of species and of habitats, cannot be compensated, and as far as possible should be avoided, partly (but not solely) for the sake of future generations.

Relatedly, E. Brown Weiss has reasonably argued that significant harm and risk to future generations should only be introduced if international agreements are in place to assign liability and provide compensation, and if the present generation can thus meet the full costs of adequately protecting its successors through such agreements and through appropriate safety mechanisms;[19] once again the near-impossibility of adequate provision for compensation implies avoiding such processes (for example, nuclear-energy generation) in the first place. With regard to nuclear-energy generation in particular, Robert E. Goodin has observed that the same verdict is delivered by a number of decision-making criteria, including those of keeping options open, risk-benefit analysis, maximin (preferring the least bad among worst possible outcomes), avoiding harm, and maximising sustainable benefits. Alternative energy strategies fare better than both nuclear and conventional energy on most of these criteria,[20] and consequentialists, concerned as they are about foreseeable benefits, risks and harms, can gladly endorse these judgements.

Weiss and Young introduce further principles on the basis of their understanding of the current generation as trustees, inheriting a common patrimony, and obliged to hand it on to successor generations in comparable condition. Adherents of the stewardship approach, defended in Chapter 3, will find these principles congenial, but they still need to be assessed in the light of the limits to the foreseeable consequences of action and of what is feasible for agents.

Thus Young suggests (on a Rawlsian basis) that each generation should leave the next generation a per capita stock no less than it inherited, and also that the total stock of resources and assimilative capacity should be maintained through time.[21] These principles, however, appear to involve costing natural resources, and to involve replacing depletions with new resources of matching value, a dubious programme where losses of species, living resources or their habitats are concerned. Even if values in pounds sterling or dollars are avoided, and nothing more than maintaining a qualitative similarity of resources per capita is envisaged, this could be prohibitively difficult to deliver granted that the human population is almost inevitably set to increase substantially for some decades. Provision of enough energy, fresh water and food for foreseeable future populations has been discussed above (Chapters 5 to 7), and may be attained if population policies and policies of sustainability are introduced; but maintaining the current ratio of resources per head could defy possibility, particularly if resources include space. Maintaining totals of renewable resources might be simple by comparison, but involves at the very least reaching a comprehensive and lasting international regime on carbon emissions, and a virtual ban on deforestation and on overfishing; at least this overall project is highly desirable and probably forms the best route to sustainability, but it may prove a daunting task for generations of politicians and negotiators to come, and in face of the existing range of perspectives (see above) will also involve widespread changes of attitude before it becomes seriously possible.

Some related principles have been suggested by Weiss, developing ideas of John Locke. Each generation is entitled to receive and obliged to maintain and transmit 'a planet-wide resource and cultural base' 'of comparable quality and diversity', providing 'each generation with similar options' and equal 'opportunities of access to the legacy from the past'. Also 'improvements made by previous generations must be conserved for all future generations', and if 'one generation fails to conserve the planet at the level of quality received, succeeding generations have an obligation to repair this damage', sharing the costs across several generations if necessary.[22] Here the same problems arise again for maintaining and transmitting a base of comparable quality, and for obligations actually to provide similar options and equal opportunities (as opposed to making this possible). Besides, these principles would require the preservation of all obsolete technology (as opposed to the kind of selections of 'the legacy of the past' offered by museums and libraries, and by World Heritage sites, about which the claims of Weiss are supportable). Nor should all 'improvements' be preserved, granted that some, such as large dams, have a limited usefulness; and as for repairing past environmental damage, which might include reclaiming much of the Sahara for agriculture, this may well assume a lower priority in the light of new problems than stemming the poleward movement of deserts generated by global warming, or preventing the possible inundation of the Maldives and of the coastal plains of Bangladesh. Trusteeship, then, must have full regard to facts, trends and foreseeable consequences; the time and energy of trustees are best devoted to satisfying basic needs for foreseeable human and non-human populations, cultural legacies and opportunities being preserved insofar as they cohere with this goal.

At the same time, as mentioned in Chapter 6, no generation can be expected to bear a disproportionate burden of obligation (although this does not justify discounting future interests to oblige the present). Such expectations would be counterproductive to the projects of maximising overall well-being, of trusteeship and of sustainability alike. For example, to expect the current generation of Indians and Chinese to curtail carbon emissions required for basic needs would infringe this principle and thus be unreasonable, besides conflicting with the requirements of equity within the current generation.

Another requirement of intergenerational equity, as Young points out, is avoidance of certain kinds of irreversible change. Most actions are irreversible under one description or another, and, as Young recognises, human ingenuity can sometimes offset or compensate for irreversible decisions. But we should still avoid irreversible change to 'essential ecological functions and processes',[23] of which the protective role of the ozone layer would be a good example. Other forms of irreversible change likely to have serious environmental consequences could reasonably be added to kinds which should be avoided, such as the destruction of an entire ecosystem vital for biodiversity, such as a tropical rainforest.

Irreversibility also figures in the Precautionary Principle, recognition of which Young makes a rule of intergenerational equity.[24] Since this Principle has a wider scope than irreversible decisions and (even) than intergenerational equity, it will be discussed in a separate section.

THE PRECAUTIONARY PRINCIPLE

The Precautionary Principle concerns the avoiding of harm that is either irreversible, or serious and reversible, but only with great difficulty and great effort. In theory it applies to cultural as well as to natural resources, but because cultural resources can often be reconstructed, and irreversibility is much more obviously a feature of environmental resources, it is to these that the Principle is mainly applied.[25] Both the irreversibility and the seriousness of harm or damage can be measured by the impact of decisions or policies on future generations, but impacts on the present and on coming decades are also relevant, these being the decades in which irretrievable or significant loss will not only be suffered but actually perceived as such, too.

The Principle declares that where there are threats of serious or irreversible damage (environmental damage included), lack of full scientific certainty or knowledge should not be used as a reason for postponing measures to prevent this damage.[26] It thus transcends principles which seek to prevent damage once a risk has been established, and concerns cases of uncertainty, where the probability of damage cannot be predicted, but where there is reason to believe it is possible. In one form or another, this Principle was adopted by German governments in the early 1980s, and has more recently been incorporated in European Union legislation and in the Maastricht Treaty.[27]

Some philosophers of science would suggest that the Principle is null and void because science never attains certainty, and as its findings, all being falsifiable in principle, do not comply with traditional understandings of knowledge either. But since the work of Karl Popper, the concept of scientific knowledge has been understood to incorporate the possibility of being shown to be false,[28] and to apply, roughly, to cognitive claims consistent with the evidence and with other unfalsified claims, and hitherto resistant to falsification itself. While controversies in this area continue, the view that there is no scientific knowledge would usually be regarded as unduly sceptical. To this it should be added that at any given time, the community of practising scientists often attains a consensus, and that for practical purposes the state of science consists in this consensus, rather than in whether any individual scientist anywhere has attained knowledge, let alone certainty. Read in this light, the Principle conveys that action to prevent irreversible or serious damage should be undertaken even in the absence of a scientific consensus on the matter.

But a willingness to act on this basis has proved crucial in the recent history of environmental decision-making. Thus the Vienna (1985) and Montreal (1987) conferences took place at a stage when there was reason to take anthropogenic ozone-depletion seriously, but before the evidence of Arctic as well as Antarctic depletion was clear; the available observations could have betokened nothing more than one area of depletion, which seemed capable of other causes than CFCs.[29] But waiting for scientific consensus could (as it turns out) have had catastrophic consequences; and the perceived possibility of this sufficed to trigger action, with provision for modification if the measures adopted turned out to be too weak or too strong. The introduction of a regulatory regime secured valuable time, and made stronger action much more feasible in 1990, after scientific consensus had been reached. There again, the Kyoto agreement owes some amount to the Precautionary Principle. For despite the consensus about the anthropogenic nature of global warming achieved among IPCC scientists (see Chapter 5), this view continues to be questioned by other scientists.[30] Yet the resulting Treaty was (happily) being signed for the European Union on the day these words were first drafted (29 April 1998).

In the context of the Precautionary Principle, Young commends conducting environmental impact assessments,[31] and (despite the danger of spurious precision implicit in conventional cost-benefit analysis) these have their place in disentangling uncertainties and in distinguishing serious damage from lesser damage. As I have argued elsewhere, such assessments have a valuable part to play, particularly if restructured so that future interests and non-human interests are taken into account.[32] But the Precautionary Principle would normally need to be invoked before such assessments could be carried through, and their current relevance could well lie in a subsequent review of precautionary decisions, rather than in advance of them.

Weaker forms of the Principle require the use of the Best Available Technology (BAT) where there is danger of transgressing the critical load of a system. Young suggests that this is justified when the cost of degradation would be serious but the degradation itself reversible, adding that a large safety margin is appropriate.[33] But BAT could easily fail to be good enough, if an environmental system is in danger of critical strain; the competence of the Best Available Technology would be fortuitous. This is all the more important where BAT is qualified with NEEC (Not Entailing Excessive Cost), since the adequacy of the best non-costly technology would be a stroke of even greater good fortune. Qualifications of the Precautionary Principle such as the conjunction of BAT and NEEC effectively undermine the Principle which they qualify; and, while there is sometimes a place for BAT, that too could undermine the Principle if the meaning of its introduction is that serious or irreversible environmental damage can now be risked.

While risk-taking must sometimes be justified, particularly where inaction could be catastrophic, nevertheless the Precautionary Principle (as presented above) is almost invariably a reliable (albeit derivative) principle, justified by the likely strongly favourable balance of good over bad consequences implicit in habitually following it. Further, since environmental systems are often endangered by action rather than inaction, the Principle would often be applicable in cases where the right policy is inaction. Each legislature should accordingly institutionalise the Precautionary Principle so as to provide for preventative injunctions and preventative interventions alike.

STRATEGIES FOR SUSTAINABILITY

The above principles comprise a combination of institutionalised caution and prudence where environmental disruption is at risk with a radical overhaul of international economic arrangements, particularly with regard to rectifying environmental degradation which has taken place already. All this is compatible with a large variety of kinds of sustainable society emerging; for, as Ted Benton has written, there need not be any single line of sustainable development, as opposed to a plurality of approaches.[34] Sustainability being the capacity for being maintained indefinitely, this observation well fits what reflection on this concept would lead us to expect.

As is argued in *Caring for the Earth* (a report of, among others, the United Nations Environment Programme), there is potential for sustainability in natural systems such as wilderness, modified systems such as naturally regenerating rangeland, cultivated systems such as farmland and even built systems such as cities, docks and railways. Even such built systems are capable of sustainability if developed in ways sensitive to both human and ecological communities. Only degraded areas which have undergone substantial pollution and lost their soil and most of their species have no potential for sustainability short of restoration or rehabilitation.[35]

Strategies for attaining sustainability in a given society include the steps which *Caring for the Earth* commends. Crises aside, extensive consultation and consensus building, plus careful analysis of all available information, should precede the adoption of policies and structures and their implementation.[36] However, the regional and global context is likely to present constraints, only capable of being overcome through international agreement and concerted action. Since sustainability requires patterns of production and consumption capable of being maintained over a long period, international agreements and policies also have to be capable of providing a stable context for sustainable development, averting (for example) disasters such as the flooding of all low-lying coastal areas. Besides, no society, however internally sustainable it may seem, deserves the accolade of sustainability if it is incompatible with a sustainable world system. Thus planning for sustainability sooner or later turns not only on domestic arrangements but also on securing agreement on principles of the general kind discussed earlier in this chapter.

But even agreement on such principles is not enough. In the final chapter, attention will be given to the kinds of decision-making which make the implementation of such principles possible, and also to the kind of civil society and of global citizenship capable of developing promising aspects of global society, overcoming less promising aspects and of fostering in each generation ways of approaching an equitable and sustainable world society.

NOTES

1. Steven Yearley, *Sociology, Environmentalism, Globalization: Reinventing the Globe*, pp. 81, 102–7.
2. Anil Agarwal and Sunita Narain, *Global Warming in an Unequal World: A Case of Environmental Colonialism*, p. 5, cited by Yearley at pp. 105–6. For some implications for international justice, see Henry Shue, 'The Unavoidability of Justice', in Andrew Hurrell and Benedict Kingsbury (eds), *The International Politics of the Environment*, 373–97.
3. Yearley, *Reinventing the Globe*, p. 106.
4. Ibid., pp. 106–7.
5. Ibid., pp. 107–9.
6. Ibid., pp. 110–13.
7. Ibid., pp. 113–15.
8. Ibid., p. 110.
9. Ibid., pp. 110–15.
10. Charles R. Beitz, *Political Theory and International Relations*, p. 141.
11. Michael Jacobs, *The Green Economy: Environment, Sustainable Development and the Politics of the Future*, pp. 182–3.
12. Jacobs, *The Green Economy*, p. 182. See also Robin Attfield, 'Unto the Third and Fourth Generations', *Second Order: An African Journal of Philosophy*, VIII.1 and 2, 1979, 55–70.
13. Jacobs, *The Green Economy*, p. 183.
14. Ibid., pp. 183–4.
15. Ibid., pp. 187–8.
16. Michael D. Young, *For Our Children's Children: Some Practical Implications of Inter-Generational Equity and the Precautionary Principle*, p. 7.
17. Ibid., p. 53.
18. Talbot Page, 'Intergenerational Justice as Opportunity', in Douglas MacLean and Peter G. Brown (eds), *Energy and the Future*, 38–58.
19. E. Brown Weiss, *In Fairness to Future Generations: International Property Law, Common Patrimony and Intergenerational Equity*, p. 80; Young, *For Our Children's Children*, p. 53.
20. Robert E. Goodin, 'Ethical Principles for Environmental Protection', in Robert Elliot and Arran Gare (eds), *Environmental Philosophy*, 3–20
21. Young, *For Our Children's Children*, p. 49.
22. Weiss, *In Fairness to Future Generations*, pp. 24–6; Young, *For Our Children's Children*, pp. 50–1.
23. Young, *For Our Children's Children*, pp. 8–9.
24. Ibid., p. 12.
25. Ibid., p. 14.
26. Ibid., p. 14.
27. Ibid., p. 14.
28. Karl Popper, *Conjectures and Refutations*.

29. Yearley, *Reinventing the Globe*, p. 111.
30. Robin McKie, 'Man "Not to Blame" for Global Warming', and 'Solar Wind Blows Away Theories', *The Observer*, 12 April 1998, pp. 1 and 9.
31. Young, *For Our Children's Children*, p. 13.
32. See Robin Attfield and Katharine Dell (eds), *Values, Conflict and the Environment*, pp. 134–8.
33. Young, *For Our Children's Children*, p. 16; see also Robin Attfield, 'The Precautionary Principle and Moral Values', in Tim O'Riordan and James Cameron (eds), *Interpreting the Precautionary Principle*, 152–64.
34. Ted Benton, 'Biology and Social Theory in the Environment Debate', in Michael Redclift and Ted Benton (eds), *Social Theory and the Global Environment*, 28–50, p. 44.
35. International Union for the Conservation of Nature, *Caring for the Earth: A Strategy for Sustainable Living*, p. 34.
36. Ibid., p. 204.

CHAPTER 11

WORLD CITIZENSHIP IN A PRECARIOUS WORLD

THE AGENT AS CAMPAIGNER

What does the ethic defended in this book (biocentric consequentialism) ask of the individual, and does it ask too much? These questions have political overtones, and this chapter delves into some of their political implications. It also explores the notions of global citizenship and of civil society; for the realities corresponding to these notions turn out to be significant if global problems are to be tackled and if individuals are to be capable of playing a part in tackling them.

A recurrent objection to consequentialist theories consists in their alleged overdemandingness, seemingly requiring the individual to become worn out optimising the balance of good over evil without respite. As this objection can be directed at several other forms of cosmopolitanism too, including Kantianism and including those approaches for which rights are fundamental, it could even be employed in the cause of alleging the superiority of egoism or of communitarianism. But this objection presupposes that consequentialism (and generally cosmopolitanism) cannot recognise any upper limit to obligation, a presupposition which I have challenged elsewhere.[1] Further, a reply was offered in Chapters 6 and 10 to the version of this objection that concerns the possible overburdening of a particular generation: it would be counterproductive, it was replied, to expect any one generation to bear a disproportionate burden of obligation. A version of this reply holds good at the level of individual agents too. However, a different reply again is also relevant here.

For, as Paul Gomberg has argued, the above objection assumes one particular model of agency, that of the individual as philanthropist, surrounded by a sea of misery, and shouldering in isolation the unending humanitarian burden of relieving or alleviating this misery. But this objection ignores other, often more accessible, models of agency, including the political model. In this model, the agent co-operates with others to change the world, or rather some of its structures, and to provide for the alleviation of misery or environmental threats or degradation through a collaborative approach, rather than through lonely and unlimited self-sacrifice. (Comparatively few agents, we can reasonably assume, will have the money or power to adopt a philanthropic role on a regular basis.) Thus to assume the philanthropic model is to invent a problem through a misleading kind of abstraction, for actual agents are free to adopt the model of campaigner instead, a model that is much more rewarding and psychologically self-sustaining.[2] Campaigning can be challenging too, but can also be enjoyable, and can foster mutual respect and self-respect among campaigners. In any case it is not intrinsically characterised by overdemandingness, as the objection predicts for ways of life in which consequentialism is consistently embodied and put into practice, for the obligation to optimise foreseeable outcomes is shared, rather than shouldered alone. (In some circumstances campaigning can be combined with witnessing against particular objectionable products through symbolic abstentions or boycotts; but witnessing should probably not be regarded as a satisfactory independent model of agency, as it would often be ineffective in the absence of campaigning.)

However, the viability of the political model of agency is sometimes held to involve preconditions far removed from those of the real world, particularly with regard to environmental campaigning. This claim, which, if upheld, would once again plunge attempts to apply biocentric consequentialism into futility, will be considered in the coming section.

DECISION-MAKING IN NON-IDEAL SITUATIONS

It is sometimes held that only societies with special decision-making procedures offer any genuine prospect of environmental sensitivity. More specifically, John S. Dryzek maintains that in an 'open society' (roughly, a community resembling Karl Popper's society of problem-solvers, 'governed by free and open conjecture and criticism'),[3] 'nature *can only* be manipulated and engineered'.[4] While Dryzek does not believe that the open society actually exists,[5] his criticism is reiterable for all forms of non-primitive society except those embodying 'communicative rationality', in which alone 'there is no obstacle to' a 'symbiotic orientation toward the natural world'.[6] In all other kinds of society, the prospects for such a positive orientation are seemingly hopeless.

A society embodying communicative rationality is defined as one approximating to Jürgen Habermas' 'ideal speech situation', where decisions are made on the basis of what would be agreed at the end of full debate in the absence of irrelevant obstacles to rationality such as differences of power.[7] Even if there never was or will be an ideal speech situation, we can employ the thought of such a situation to assess the fairness and validity of decision-making. By contrast situations where decisions are affected by arbitrary factors are certain to be exploitative. While Habermas himself regards the natural world as comprising objects rather than subjects, and as unfit to enter into situations of speech, Dryzek hopes that ecosystems, as self-regulating systems, can send feedback signals to speaking subjects and 'tell' of the interests which need to be heeded.[8] If value belongs only to entities capable of entering into communication, Gaia may even so be held to satisfy this requirement.[9]

Decisions should certainly be made in as fair and rational a manner as possible. But ecosystems are not speakers; it is humans that discover how far the biosphere (or 'Gaia') is self-regulating, and how ecosystems are best preserved, to the extent that they can be identified at all (see Chapter 6 above), and it is humans that are capable of reflection on the well-being of their individual living members. This has important implications for decision-making. For some interests can and should be considered which are not represented by their holders in the forums where decisions are made; the project of broadening ideal speech situations to include non-humans is unnecessary, as well as futile. Thus decision-makers should not suppose that the interests to be considered are confined to the interests of themselves (the speakers or participants) and their electorates, or that intrinsic value is confined to beings with a voice. For if decisions are based on the interests of those represented directly or indirectly in such forums, then the well-being of entire sectors of affected beings is liable to be neglected. In these circumstances, decision-making is liable to be anthropocentric, as Robyn Eckersley argues,[10] and even within humanity, future generations are likely to be ignored. Since all this happens all too often, the implications should be pursued further.

Before procedures are suggested for avoiding this neglect, Dryzek's appraisal of open societies should be addressed, namely that nature is certain to be manipulated in such societies, and therefore also in those actual societies (such as representative democracies) which share some of the virtues of an open society (such as freedom of speech and inquiry) but are in some regards compromised by lack of openness. Unsurprisingly, the well-being of non-human nature and of future generations is frequently neglected in such societies, at least as much as it would be in an ideal speech situation; so the prediction that nature will sometimes be manipulated is frequently upheld. However, the implicit prediction that nature will invariably be manipulated in liberal-democratic societies is not borne out by the facts; for example, wild species are sometimes protected, cruelty (and neglect) towards individual creatures is widely forbidden, and these policies and the related legislation are sometimes enforced. So blanket pessimism on the part of biocentric (or of ecocentric) people is out of place with regard to representative democracies, despite the distorting influence of commercial interests there (and is thus out of place with regard to less compromised open societies too). Whether blanket pessimism is in place about any society is an issue to which I will return.

There are, however, stronger grounds for optimism regarding ecological sensitivity where all sections of society are involved in decision-making. Only in this way are people likely to take full responsibility for their dealings with nature and their use of natural resources. Some writers even include the democratic participation of all sections of society in definitions of sustainability;[11] but it is unwise to stipulate the presence of everything desirable, including participatory democracy, before a society can be called 'sustainable', as a less demanding approach makes sustainability less out of reach and the concept more useful in practice.[12] Yet there can be little doubt that participatory democracy and sustainability are normally well adapted to each other – a partial vindication, this, of Dryzek's views on decision-making procedures.

However, the question arises of how to take into account interests not represented even in ideal speech situations, or societies modelled on them, and, more importantly, how to do so in representative democracies which do not represent these interests either, particularly where these interests are not represented in any other national legislature either. The interests of non-human nature and of future generations are unlikely to be adequately represented unless human proxies are appointed to legislatures, charged with representing these interests,[13] and supported by research staff and facilities. Since ideally all legislators would heed these interests, these additional legislators should not be introduced in such numerical strength as to allow the others to claim that the constituencies of nature and of future generations no longer need their own attention; otherwise there would be a case for the newcomers to match or even outnumber existing legislators so as to reflect the large numbers of (for example) future people likely to exist in due course. The role of these proxy legislators would rather be to represent otherwise absent interests, and to persuade their colleagues to take them seriously.

The case for this change is a serious one. On the one hand, moral standing has been argued to extend to future generations and to non-human living creatures; on the other hand, existing legislatures are structured to disregard these interests. While it would be wrong to seek to enforce morality through legislation, there should not be such a large structural gap between legislatures and what is sometimes called 'the moral constituency'; and the suggested change is probably the most streamlined and least disruptive way of remedying the deficiency. There again, if the model of human stewardship or trusteeship of the planet is accepted (see Chapter 3), then some institutionalised provision is required to carry through the associated responsibilities; and the proposed innovation is a minimal form which this provision could take. Thus the proxies would be likely to take a special interest in the avoidance of global anthropogenic catastrophes. Admittedly, the proxies appointed could not be answerable to the interests which they would be appointed to represent; but this apparent problem could be addressed by providing for them to be appointed by, and to answer for the discharge of their duties to, as representative a body as could be devised, granted the nature of the interests in question. The largest objection to what is here proposed is that it involves a dilution of democracy; but if unmodified democracy is unlikely to provide for interests which ought to be provided for, but can do this much better in the manner proposed, and if its central structures and virtues can at the same time be maintained (as they undoubtedly can), then this is not an objection strong enough to be allowed to stand in the way of the proposed change.

There is a more general implication to draw from the possibilities for ecological sensitivity (and for global solidarity) which exist in imperfect representative democracies already, and which could be enhanced along the lines of the proposal just introduced. This is that societies and situations which fall short of being ideal for the introduction of environmentally sensitive actions and policies should not be regarded as uniformly hopeless, nor regarded with blanket pessimism. The suggestion that nothing can change for the better until all society's structures have been positively transformed, which metimes underlies the contrary belief, is actually a counsel of premature and unnecessary despair. Admittedly some crisis situations make all considerations other than the struggle for immediate survival distractions from that struggle, and in such circumstances people's responsibilities with regard to future generations or non-human creatures lapse, except insofar as they support one of the options leading to survival. But that still leaves numerous less-than-ideal situations in which individual agents, families or groups have some limited scope for choice, or for introducing practices or precedents conducive to ecological sensitivity or social justice or global solidarity. Billions of humans are subject to authoritarian regimes, but are not completely deprived thereby of all opportunities for forming constructive decisions in their daily lives. Hence, while ideal decision-making situations that philosophers or other theorists describe frequently remain unachievable, this cannot be allowed to absolve the majority of humanity living in non-ideal conditions from all responsibility, much less from environmental responsibility in particular. Greater responsibility, however, attaches to people living in relatively open societies and with greater-than-average opportunities.

GLOBALISATION AND RELATED CONCEPTS OF GLOBAL CITIZENSHIP

The role of people aware of global problems and of their own responsibility for ameliorating them can best be understood as one of global citizenship. Several themes of this book can be related to a contemporary understanding of global citizenship, including those of the shared global environment, of our shared trusteeship of the planet, and of shared responsibilities for justice between contemporaries, between species and between generations. A global ethic, it may reasonably he held, requires us to think of ourselves as global citizens. But first the relation should be explored between one or another conception of global citizenship and the phenomenon of globalisation.

'Globalisation' refers to the process of worldwide economic and cultural integration. More particularly, it concerns the integration of financial markets, the increasing power and outreach of international corporations, an increasing worldwide uniformity of style of places like hotels and airports, increasing international communications through technological innovations such as the internet, and the increasing use of English as an international language. The liberalisation and deregulation of trade are characteristic features of this process, which promises (or threatens) to open up to world business the most remote fastnesses on Earth.

Globalisation should not be regarded as uniformly bad (for example, the internet allows people with ethical concerns in different countries to communicate and co-ordinate more readily) nor as typically good (for example, it is probably responsible for the accelerated destruction of numerous forests, and undoubtedly responsible for numerous harmful Structural Adjustment Programmes). Nor should developments of a globalising kind be regarded as inevitable, or their characteristic ideology as irresistible, or protest against them as hopeless, in view of well-documented cases of successful resistance, and generally of human creativity and ingenuity.[14] Globalisation is no more the end of history than feudalism was. But it does comprise a significant element of the scenery of contemporary life, and thus of the context of global citizenship for the opening decades of the new millennium. Further, calls to obliterate and replace it with decentralised, local structures can be misguided, in view of the need for global structures to tackle global problems.

Several conceptions of global citizenship should be distinguished, some of them closely related to globalisation. One conception assumes a strong notion of citizenship, for which a citizen necessarily bears the rights and duties of a member of a state which is politically sovereign. Hence global citizenship could only be fully present if the present plurality of states were replaced by a sovereign global state. As was observed in Chapter 9, global problems (global environmental problems not least) are sometimes perceived as requiring nothing less than such world government, possibly through a pooling of sovereignty in the United Nations Organisation. Advocates of such political restructuring could thus be regarded as adherents of an institutional conception of global citizenship, a form of global citizenship depicted by Richard Falk as essentially focused on structural change at global level.[15] Supporters of this view could regard globalisation as strengthening their case, since a global sovereign would be in a stronger position than less powerful structures of international co-operation to constrain or regulate the global market, and to tax it to compensate its victims. Indeed Hobbesian realists, as Janna Thompson has remarked, could add their support to this view as the 'only viable means for imposing environmental restrictions on a reluctant world'.[16] But as was also argued in Chapter 9, there are some apparently conclusive objections to world government; and in the context of globalisation it could be added that a world government would be unlikely to prove the best structure for securing local consent for international policies of constraint in resource-usage or resisting the homogenising tendencies of globalisation or preserving cultural diversity. The United Nations is undoubtedly crucial to the future of humanity and the planet, but world sovereignty need not (and probably should not) be its future role.

Hence it is important that there are other conceptions of global citizenship. In a second conception mentioned by Falk, citizenship is conceived neither in terms of political affiliation nor of local loyalty, but of participation in a global culture, such as the culture of global business. Falk depicts a businessman met on an aeroplane, who claimed to be a global citizen on the counts of his global travels and global network, of staying at the same kinds of hotels whatever country he happened to be visiting, and of having discarded meaningful local loyalties. Global citizenship as membership of a 'homogenised elite global culture'[17] thus involves being an agent of globalisation; this kind of citizenship is a matter of apolitical club-membership rather than of loyalties, whether patriotic or internationalist.

The rise of globalisation is probably generating many such global citizens. But this kind of global citizenship involves neither concern about global problems and the global environment, nor a sense of trusteeship (except where trust laws obtrude), nor concern for powerless people, nor other generations nor other species. Cosmopolitans, however, as was seen in Chapter 2, need not discard local loyalties; by contrast, the second kind of global citizen seems to combine lack of local roots with lack of global concern. Nor is his a promising route to the preservation of cultural diversity. For less inadequate conceptions of global citizenship, we have to look further.

ALTERNATIVE CONCEPTIONS OF GLOBAL CITIZENSHIP

Unedifying as the second conception of global citizenship may be, it draws to attention the fact that citizenship need not involve loyalty to an existing political sovereign. Citizens can also be conceived as members of a community that is bound together by characteristics other than a political structure; indeed this is probably a commoner sense of 'community' than the political sense. The issue about global citizenship now turns out to be whether communities are essentially either sectional or local (as presumably some communitarians believe), or whether in a significant sense there can be citizenship of a global community.

Belief in such a kind of global citizenship was held by some ancient Stoics, as was remarked in Chapter 2, and without forfeiture either of local identification or of wider concerns about social issues such as hunger, as has been well brought out by Martha Nussbaum.[18] In more recent times, this notion has been resuscitated, among others by Kant (see Chapter 2); for the duty of rational agents to work towards a confederation of free states neither discards local loyalties nor seeks to replace local states with a world state, as opposed to a cosmopolitan world order.[19] For his part, Falk introduces no less than three further models of global citizenship, each embodying loyalties wider than national loyalties without a focus on world government. His third conception of global citizenship is focused on global management, particularly with regard to ecological and economic dimensions; a central example is the approach of the Brundtland Report, with its advocacy of unprecedented international co-operation to tackle urgent problems of environment and development.[20] His fourth conception is exemplified in the new consciousness, among countries of the European Union, of a European identity and the merits of European integration, but he recognises that such regionalism, despite its global benefits, could form a potential obstacle to fully global co-operation.[21] His fifth and final conception comprises the global citizenship of transnational activism; examples include Greenpeace, Amnesty International and the women's movement.[22]

For many, the European ideal has superseded the nationalistic chauvinism and belligerence of the first half of the twentieth century, and the deepening and widening of the European Union promises to form a transitional stage on the path to stronger international co-operation. In the negotiations leading up to the Kyoto agreement, for example, the European Union played a valuable role in persuading the other developed nations to join a new greenhouse gas regimen on principles acceptable to the Third World.[23] But while the European experience can serve as a precedent for overcoming traditional enmities and for a partial pooling of national sovereignty, there are also dangers of a fortress Europe approach to the rest of the world, comprising the epitome of all that is worst in Eurocentrism; such dangers are just as compatible with Europeanism as its international promise. Thus Falk's fourth conception of global citizenship is not essentially global at all.

But his third (global management) and fifth (global activism) conceptions need not be contrasted with each other. Global activists characteristically seek to reform the principles, policies or performance of international bodies such as the United Nations, the World Bank and the International Court as much as those of national governments and of multinational corporations. In other words, improved global management is one of their objectives, embodying enhanced international observance of the Universal Declaration of Human Rights, or of the Cairo Programme of Action on women's education and reproductive rights (see Chapter 7), or of Agenda 21 (the programme of environment and development action agreed at Rio). Frustrating as they may find the global bureaucracy to be, global activists, for all their ethical concern for a better world, themselves depend on systems of global governance.

The converse, of course, does not hold, and global managers do not usually rely on NGOs (non-governmental organisations) to carry out their work, although in recent years they have found it beneficial to involve NGOs in international conferences (as at Cairo in 1994, and at Beijing, the International Conference on Women, in 1996), and to involve humanitarian NGOs in handling famine relief. But there is in any case no good reason for Falk's third conception to be tied to the perspectives of global managers as an interest-group, let alone to the tendentious perspective which he proceeds to present. Falk muddies the waters when he declares that 'What it means to think of global citizenship from this functional perspective is increasingly caught up in the process of making the planet sustainable at current middle-class life-styles, which means making the carrying capacity of the planet fit what happens in different parts of the world.' Such remarks enable him to generate contrasts with his fourth and fifth conceptions, but they also generate contradictions. The supporters of *Our Common Future*, of Agenda 21 (which he mentions in the same breath) and of planetary sustainability certainly advocate unprecedented international co-operation and managerial effort at international level, but they would be inconsistent, and committed to undermining their own projects, if they aimed to globalise and perpetuate current middle-class lifestyles including current levels of consumption of energy, water and other resources, or if they sought somehow to adapt the carrying capacity of the planet (by interplanetary trading, perhaps?) to this level of resource-use. Even if the perspective presented here by Falk is held by some global managers, it is actually incompatible with the principles for global management presented by Brundtland, in Agenda 21 and in the Rio Declaration and Conventions. While these documents are not flawless, they are not significantly flawed by managerialism or elitism.

Thus supporters of these documents and declarations can, as Falk suggests, be regarded as answering to a conception of global citizenship involving global management; but answering to this conception need not imply sharing in the managerialism of elitist international bureaucrats, contrary to what Falk implies. Such supporters will sometimes be found as transnational activists, for example as members or staff of environmental NGOs, sometimes as local campaigners, such as members of the United National Association, and writers of letters to their elected representatives on international issues, and sometimes as managers themselves, whether at national or international level. Their global citizenship consists in recognition of cosmopolitan principles and loyalties, applied to global as well as local issues, and in attempts to apply them in practice. Understood in this light, Falk's third and fifth conceptions of global citizenship, involving respectively support for global management and for global activism, merge into a single, distinctive and commendable conception. This global citizenship will often involve participation in citizens' networks (and sometimes in networks of global citizens), which are sometimes held to be necessary for democracy, and arguably necessary also if global environmental problems are to be consistently tackled with sufficient urgency. For, given the global ethic defended in this book, the sphere of concern of global citizens cannot be regarded as confined to human fellow citizens and their successors, but extends to all present and future fellow creatures of the planet.

CIVIL SOCIETY

Citizens' networks comprise an important part of civil society. Civil society, in current usage, incorporates associations independent of the state,[24] such as trade unions, churches, NGOs, political parties and societies for neighbourhoods, hobbies or sports. 'Civil society' can also refer to the market,[25] and this helps explain the critique of the later Marx of civil society as exploitative;[26] indeed where civil society, broadly conceived, involves systemic evils such as racial discrimination, class or caste distinction and the oppression of women, this is a reasonable verdict. But a different verdict is in place, as O'Neill remarks, where 'civil society' primarily refers to voluntary associations and networks;[27] indeed the presence of these in a country is often thought to be necessary for the maintenance of democracy there.[28]

Just as good procedures and wise decision-making are plausibly facilitated by a vigorous civil society at local and national level, global issues are much more likely to be properly addressed and tackled only when global civil society is in place, at least embryonically, ventilating problems and lobbying for constructive solutions. In Marx's sense, for which global civil society includes global markets, civil society is part of the problem at this level too, but in the sense of international networks of voluntary NGOs and their individual members, it is frequently an ingredient of possible solutions. In this sense, international civil society partly comprises committed international non-governmental bodies like Greenpeace and Amnesty International, or like the International Red Cross, Médecins sans Frontières and the International Campaign to Ban Land Mines.

It also includes such non-governmental international bodies as the World Council of Churches, the Roman Catholic Church, and other worldwide religious bodies, or again the International Soccer Federation,[29] for bodies can contribute to global civil society without embracing the preservation of the global environment among their aims. International peace groups, federations of trade unions, and networks of organisations of indigenous people, minorities and ethnic groups (such as the Environmental Justice Movement) should not be forgotten. But it is also comprised by individual citizens and pressure-groups, and the networks in which they participate, including those harbouring concerns for environments other than their own, such as the Antarctic or the Amazon Basin (or, as mentioned in Chapter 1, for the global environment); fortunately, as Thompson points out, people do develop such concerns, and without this the prospects for the global environment might well be slender.[30] Campaigning focuses not only on international conferences (such as Kyoto) and organisations (like IMF) but also on transnational corporations, as when the World Development Movement (WDM) persuaded Del Monte to allow independent trade unions in its banana plantations in Central America, and as in WDM's recent lobbying of other large banana companies to do likewise, and to curtail the aerial spraying of bananas, the environment and plantation workers alike with herbicides and pesticides.

Campaigning NGOs cannot be assumed to be flawless, sometimes oppose each other, and are sometimes open to significant criticism, as over the failure of the feminist lobby at the Cairo Population

Conference to stress the structural requirements of development (see Chapter 7). But more often they work together, as in the Real World Alliance in Britain;[31] and they probably represent the best hope of kindling willing compliance with what the Commission on Global Governance sensibly claims to be vital to good global governance, a global civic ethic.[32] Rights and responsibilities are proposed by the Commission in this connection for all members of humanity. The details of its (rather anthropocentric) lists cannot be debated here, but it is significant that the responsibilities include 'consideration of the impact of . . . actions on the security and welfare of others' and 'protection of the interests of future generations by pursuing sustainable development and safeguarding the global commons'.[33] The Commission rightly applies this ethic not only to governments and NGOs, but also to transnationals, academia and the mass media.[34] International civil society (and the global citizens who comprise it) has, I suggest, a substantial capacity to affect the degree to which power is wielded ethically in all these bodies.

The Commission includes in its list of human rights a right of equal access to the global commons,[35] recognition of which (as was seen in Chapter 5) could have radical implications for carbon emissions regimes. As Tim Hayward relates, recent European constitutions have included declarations of citizens' environmental rights, and as he argues, the further spread of this practice when constitutions are drafted and ratified could well make the pursuit of sustainable policies harder to destabilise, and thus itself more sustainable.[36] This is one among many instruments usable to enhance the prospects of a sustainable world society, committed to a strong global ethic, and also a sustainable world.

Meanwhile the widespread adoption of a global ethic along the lines commended by the Commission, or along the general lines of the global ethic of Hans Küng and Karl-Josef Kuschel (see Chapter 1), preferably in a version with a biocentric and consequentialist basis such as that advocated in this book, could help to solve the problem of motivation, that is, the widespread reluctance of individuals and communities to accept self-restraint or, in some cases, to welcome (or at least tolerate) change for the sake of the planetary environment, of future generations or of the common good. For the adoption of such an ethic, whether or not associated with a matching metaphysic (see below), can confer the kind of self-transcendence and sense of meaningfulness noted in Chapter 4 (above) as a widespread psychological need; once such an ethic is shared among fellow members of a society, or of a campaigning group, it can also (as mentioned there) become psychologically self-reinforcing.[37] The sharing of an ethic of environmental concern is important for sustainability, at least at a local level; communities willing to take responsibility for their shared environment are also much more likely to care about the global environment than others, even if parochial or local concern is not the only route to commitment at the global level. However, the sharing of a global ethic that crosses spatial and temporal boundaries is no less important, together with attitudes or motivations such as identification with fellow-people or fellow-creatures, a sense of awe or humility before nature or creation, or a sense of global justice. This ethic also needs to be institutionalised in networks and systems of co-operation, such as the United Nations family of organisations, without which it could be rendered fruitless. But this is not the place for further discussion of political systems for environmental decision-making. Systems are in any case inert without ethical values, and without attitudes such as love of what is valuable.

FURTHER GENERAL CONCLUSIONS

Many of the conclusions adopted in the preceding chapters (and some of those of earlier sections of this chapter) could be adopted on either a realist basis or a communitarian basis. Consider first some possible realist grounds for accepting national quotas for carbon emissions. This acceptance could be a matter of enlightened self-interest, if enough people were prepared to vote for it to make supporting it politically popular, or if there were profits to be made for local industry from an early switch to energy-efficient technology, which quotas would make more profitable. Or there could be allies to be made and business deals to be secured through early support of bodies like the Association of Small Island States if Southern states with larger markets were thought likely to champion its cause. (But note how most of these grounds depend on others recognising the intrinsic merits of the case for carbon-emission quotas, as opposed to contingent advantages.) Similarly realists could endorse biodiversity conservation, whether on grounds of votes or business or alliances. However, they could just as easily reject environmental principles and policies on the same grounds, if votes or business or alliances so indicated. Thus a better foundation for global sustainability and global justice is needed. However, the very flexibility of egoistic reasoning to the values of its social environment underwrites the importance of proponents of less shallow kinds of ethic spreading their values as vigorously as possible. For egoists are always likely to be on hand, adapting their direction to the way the wind is blowing.

Communitarianism too can endorse (for example) principles of intergenerational equity, as de-Shalit has ingeniously shown, as long as this endorsement expresses a community's current values.[38] Such endorsement could readily be extended to support for Third World development, population policies, and other policies commended here, and is to be welcomed. But, as Dower has observed, such endorsements are entirely contingent on community values being and remaining more or less what they are; in the absence of these valuations, the related obligations would not only forfeit recognition, but would become groundless, or, from the communitarian perspective, non-existent.[39] If, as I have been maintaining and as most people believe, agents' obligations remain in place even when the values held by those agents cease to affirm them, then communitarian theories of their grounds and status are inadequate. As was argued in Chapter 8 with relation to preservation, communitarian arguments, despite their local or temporary resonance, are no substitute for arguments of a more pervasive and perennial, cosmopolitan kind. (These considerations could also be harnessed to an argument for meta-ethical objectivism, which, however, falls beyond the scope of this book; I have presented such an argument elsewhere.)[40]

Among kinds of normative ethics, only cosmopolitanism does justice to the objective importance of all agents heeding ethical reasons, insofar as they have scope for choice and control over their actions, and working towards a just and sustainable world society. The universal values that it upholds include many of the values in which communities and nations take pride (in Britain, values such as freedom, tolerance and fair play), for these are implicitly universal and cosmopolitan values, for which communitarians can claim no monopoly. And among cosmopolitan approaches, as I have argued at several points above, only consequentialist theories take future needs (including the needs of our possible successors) adequately into account, at the same time as those of contemporaries. (Consequentialism based on needs should be carefully distinguished here from utilitarianism, a quite different form of consequentialism based either on happiness or on well-being defined in terms of preferences.) And only biocentric theories do this in a manner which transcends human interests, taking account of the importance of biotic systems, but doing so without prioritising these systems over the valuable creatures which they sustain.

As Dower has pointed out, we should avoid the anthropocentric view that the resources of the planet are the common heritage of humankind alone.[41] But humanity has a related inheritance, in the form of trusteeship of the biosphere; this view was at least argued in Chapter 3 to be a coherent one, whether on a theistic basis or a secular one. Strictly this is a metaphysical view, which transcends ethics, both in its religious versions and in its secular version (for which current agents are answerable for the care of the planet to the intergenerational community of humanity). Nevertheless it readily coheres with the consequentialist ethic of global, interspecies and intergenerational equity (see Chapters 2 and 10), and with the principles of the intervening chapters, of securing the future of the human enterprise, conserving renewable resources, husbanding non-renewable resources, pursuing sustainable development, alleviating poverty, avoiding overpopulation, preserving biodiversity and promoting global justice. At the same time, by expressing the relation of humanity to the planetary biosphere, it can inspire commitment to the kind of ethic just described, and in this way supplement the reasons for action internal to that ethic.

For readers unconvinced of the metaphysic of stewardship, all the ethical reasons for the kind of global citizenship that I have been defending remain available undiminished, for the intrinsic value which is pivotal to a consequentialist ethic consists precisely in the existence of non-derivative interpersonal reasons for promoting or preserving whatever bears this value. But for those dissatisfied until offered an ampler self-understanding, or the possibility of lives of participation in a larger scene, in which humanity is related to nature and to the shared but vulnerable natural environment[42] of the planet, I commend the metaphysic (whether in theistic or secular form)[43] in which human beings are global stewards as well as global citizens, and the planetary biosphere is a trust.

NOTES

1. Robin Attfield, *Value, Obligation and Meta-Ethics*, pp. 115–19.
2. Paul Gomberg, 'Self and Others in Bentham and Sidgwick', *History of Philosophy Quarterly*, 3, 1986, 437–48. See also Attfield, *Value, Obligation and Meta-Ethics*, pp. 186–9.
3. John S. Dryzek, *Rational Ecology: Environment and Political Economy*, p. 187.
4. Ibid., p. 207.
5. Ibid., p. 187.

6. Ibid., p. 207.
7. Dryzek, *Rational Ecology*, p. 202. See also Jürgen Habermas, *Legitimation Crisis*, pp. xvii-xviii, where his translator, Thomas McCarthy, expounds Habermas' original use of this phrase.
8. Dryzek, *Rational Ecology*, pp. 206–7.
9. John S. Dryzek, 'Green Reason: Communicative Ethics for the Biosphere', in Lori Gruen and Dale Jamieson (eds), *Reflecting on Nature: Readings in Environmental Philosophy*, 159–74, p. 167.
10. Robyn Eckersley, *Environmentalism and Political Theory: Towards an Ecocentric Approach*, pp. 106–16.
11. Michael Jacobs, *The Politics of the Real World*, p. 26.
12. Robin Attfield and Barry Wilkins, 'Sustainability', *Environmental Values*, 3.2, 1994, 155–8.
13. Proposals along these lines have also been put forward in Edith Brown Weiss, *In Fairness to Future Generations: International Property Law, Common Patrimony and Intergenerational Equity*.
14. See Paul Vallely and Ian Linden, 'Doing Battle with Globalisation', 1, 'Are We Masters or Servants?', *The Tablet*, 9 August 1997, 1004–5.
15. Richard Falk, 'The Making of Global Citizenship', in Bart van Steenbergen (ed.), *The Condition of Citizenship*, 127–40, pp. 132–4.
16. Janna Thompson, 'Toward a Green World Order: Environment and World Politics' (unpublished), p. 6.
17. Falk, 'The Making of Global Citizenship', p. 134.
18. Martha Nussbaum, 'Patriotism and Cosmopolitanism', *Boston Review*, XIX.5, October/November 1994, 2–4, p. 3; also 'Asking the Right Questions: Martha Nussbaum Responds', *Boston Review*, XIX.5, October/November 1994, 33–4, p. 34.
19. Immanuel Kant, 'Perpetual Peace: A Philosophical Sketch', in Hans Reiss (ed.), *Kant: Political Writings*, 93–130.
20. Falk, 'The Making of Global Citizenship', pp. 135–6.
21. Ibid, pp. 136–8.
22. Ibid, pp. 138–9.
23. Michael Grubb and Dean Anderson (eds), *The Emerging International Regime for Climate Change: Structures and Options after Berlin*.
24. Michael Walzer, 'Introduction', in Walzer (ed.), *Toward a Global Civil Society*, 1–4, p. 1.
25. John O'Neill, *Ecology, Policy and Politics*, 1993, p. 177.
26. Barry Wilkins, 'Civil Society', in Michael Payne (ed.), *A Dictionary of Cultural and Critical Theory*, 102–3.
27. O'Neill, *Ecology, Policy and Politics*, pp. 177–8.
28. Michael Walzer, 'The Civil Society Argument', in Ronald Beiner (ed.), *Theorizing Citizenship*, 153–74, p. 170.
29. David A. Crocker, *Transitional Justice and International Civil Society* (Working Paper no. 13), p. 39.
30. Thompson, 'Toward a Green World Order', p. 10.

31. See Jacobs, *The Politics of the Real World*, written for the Real World Coalition.
32. Commission on Global Governance, *Our Global Neighbourhood*, p. 335.
33. Ibid., p. 337.
34. Ibid., p. 335.
35. Ibid., p. 336.
36. Tim Hayward, 'Constitutional Environmental Rights and Liberal Democracy', unpublished paper presented at European Consortium for Political Research, University of Warwick, 1998.
37. See Nigel Dower, *World Ethics: The New Agenda*, ch. 6.
38. Avner de-Shalit, *Why Posterity Matters: Environmental Policies and Future Generations*.
39. Dower, *World Ethics*, ch. 10.
40. Attfield, *Value, Obligation and Meta-Ethics*, chs 12–14.
41. Dower, *World Ethics*, ch. 10.
42. For an elucidation of the concepts of nature and natural environment, see Chapter 1.
43. See Chapter 3.

BIBLIOGRAPHY

Abberley, Doug (ed.) (1993), *Boundaries of Home: Mapping for Local Empowerment*, Gabriola Island and Philadelphia: New Society Publishers.

Addison, Joseph (1714), *The Spectator*, no. 583, 20 August.

Agarwal, Anil (1985), 'Ecological Destruction and the Emerging Patterns of Poverty and People's Protests in Rural India', *Social Action and Social Trends* (New Delhi), January to March, 57.

Agarwal, Anil, and Sunita Narain (1991), *Global Warming in an Unequal World: A Case of Environmental Colonialism*, Delhi: Centre for Science and Environment.

Aiken, William (1990), 'Famine and Distribution', *Journal of Philosophy*, 87, 642–3.

Aiken, William (1996), 'The "Carrying Capacity" Equivocation', in William Aiken and Hugh LaFollette (eds), *World Hunger and Morality* (2nd edn), Upper Saddle River, NJ: Prentice-Hall, 16–25.

Aiken, William, and Hugh LaFollette (eds) (1996), *World Hunger and Morality* (2nd edn), Upper Saddle River, NJ: Prentice-Hall.

Aiken, William (unpublished), 'Development and Population Policy', paper presented at 1996 Aberdeen conference on 'Ethics, Development and Global Values'.

Allison, Lincoln (1991), *Ecology and Utility: The Philosophical Dilemmas of Planetary Management*, Leicester and London: Leicester University Press.

Almond, Brenda (1990), 'Alasdair MacIntyre: The Virtue of Tradition', *Journal of Applied Philosophy*, 7, 102–3.

Anderson, Terry L., and Donald R. Leal (1992), *Free Market Environmentalism*, Boulder: Westview Press.

Arendt, Hannah (1958), *The Human Condition*, Chicago: University of Chicago Press.

Arler, Finn (1995), 'Justice in the Air: Energy Policy, Greenhouse Effect, and the Question of Global Justice', *Human Ecology Review*, 2, 40–61.

Attfield, Robin (1979), 'Unto the Third and Fourth Generations', *Second Order: An African Journal of Philosophy*, VIII.1 and 2, 55–70.

Attfield, Robin (1986), 'Development: Some Areas of Consensus', *Journal of Social Philosophy*, XVII, Summer, 36–44.

Attfield, Robin (1987), *A Theory of Value and Obligation*, London, New York and Sydney: Croom Helm.

Attfield, Robin (1991), *The Ethics of Environmental Concern* (2nd edn), Athens, GA and London: University of Georgia Press.

Attfield, Robin (1992), 'Development and Environmentalism', in Robin Attfield and Barry Wilkins (eds), *International Justice and the Third World: Essays in the Philosophy of Development*, London and New York: Routledge, 151–68.

Attfield, Robin (1994), 'The Precautionary Principle and Moral Values', in Tim O'Riordan and James Cameron (eds), *Interpreting the Precautionary Principle*, London: Cameron & May, 152–64.

Attfield, Robin (1994), *Environmental Philosophy: Principles and Prospects*, Aldershot: Avebury, and Brookfield, VT: Ashgate.

Attfield, Robin (1995), *Value, Obligation and Meta-Ethics*, Amsterdam and Atlanta, GA: Rodopi.

Attfield, Robin (1997), 'Discounting, Jamieson's Trilemma and Representing the Future', in Tim Hayward and John O'Neill (eds), *Justice, Property and the Environment: Social and Legal Perspectives*, Aldershot: Ashgate, 85–96.

Attfield, Robin (1998), 'Environmental Ethics and Intergenerational Equity', *Inquiry*, 41.2, 207–22.

Attfield, Robin (1998), 'Saving Nature, Feeding People and Ethics', *Environmental Values*, 7, 291–304.

Attfield, Robin (forthcoming), 'Christianity', in Dale Jamieson (ed.), *A Companion to Environmental Philosophy*, Oxford: Blackwell.

Attfield, Robin, and Barry Wilkins (eds) (1992), *International Justice and the Third World: Essays in the Philosophy of Development*, London and New York: Routledge.

Attfield, Robin, and Barry Wilkins (1994), 'Sustainability', *Environmental Values*, 3.2, 155–8.

Attfield, Robin, and Andrew Belsey (eds) (1994), *Philosophy and the Natural Environment*, Cambridge, New York and Melbourne: Cambridge University Press.

Attfield, Robin, and Katharine Dell (eds) (1996), *Values, Conflict and the Environment* (2nd edn), Aldershot: Avebury and Brookfield, VT: Ashgate.

Banister, Judith (1984), 'An Analysis of Recent Data on the Population of China', *Population and Development Review*, 10.1, 441–71.

Barnaby, Frank (1988), *The Gaia Peace Atlas*, London: Pan Books.

Barry, Brian (1973), *The Liberal Theory of Justice: A Critical Examination of the Principal Doctrines in A Theory of Justice by John Rawls*, Oxford: Clarendon Press.

Barry, Brian (1991), 'The Ethics of Resource Depletion', in Brian Barry,

Liberty and Justice: Essays in Political Theory 2, Oxford: Clarendon Press, 259–73.

Barry, Brian (1991), *Liberty and Justice: Essays in Political Theory 2*, Oxford: Clarendon Press.

Bauckham, Richard (1994), 'Jesus and the Wild Animals (Mark 1:13): A Christological Image for an Ecological Age', in J. B. Green and M. Turner (eds), *Jesus of Nazareth: Lord and Christ: Essays on the Historical Jesus and New Testament Christology*, Grand Rapids, MI: Eerdmans, 3–21.

Beckerman, Wilfred (1994), 'Sustainable Development: Is it a Useful Concept?', *Environmental Values*, 3.3, 191–204.

Beiner, Ronald (ed.) (1995), *Theorizing Citizenship*, Albany, NY: State University of New York Press.

Beitz, Charles R. (1979), *Political Theory and International Relations*, Princeton, NJ: Princeton University Press.

Belsey, Andrew (1994), 'Chaos and Order, Environment and Anarchy', in Robin Attfield and Andrew Belsey (eds), *Philosophy and the Natural Environment*, Cambridge, New York and Melbourne: Cambridge University Press, 157–67.

Benton, Ted (1994), 'Biology and Social Theory in the Environment Debate', in Michael Redclift and Ted Benton (eds), *Social Theory and the Global Environment*, London and New York: Routledge, 28–50.

Berry, R. J. (1995), 'Creation and the Environment', *Science and Christian Belief*, 7.1, 21–43.

Black, John (1970), *Man's Dominion: The Search for Ecological Responsibility*, Edinburgh: Edinburgh University Press.

Bookchin, Murray (1987), 'Thinking Ecologically: A Dialectical Approach', *Our Generation*, 18.2, 3–40.

Borza, Karen L., and Dale Jamieson (1990), *Global Change and Biodiversity Loss: Some Impediments to Response*, Boulder: Center for Space and Geosciences Policy, University of Colorado.

Brack, Duncan (1995), 'Balancing Trade and the Environment', *International Affairs*, 71.3, 497–514.

Bratton, Susan Power (1988), 'The Original Desert Solitaire: Early Christian Monasticism and Wilderness', *Environmental Ethics*, 10, 31–53.

Bread for the World Institute (1994), *Hunger 1995: Causes of Hunger* (Fifth Annual Report on the State of World Hunger), Silver Spring, MA: Bread for the World Institute.

Broome, John (1992), *Counting the Cost of Global Warming*, Cambridge: White Horse Press.

Brown, Lester R. et al (eds) (1996), *State of the World, 1996: A Worldwatch Institute Report on Progress Towards a Sustainable Society*, London: Earthscan.

Brown, Lester R., and Erik Eckholm (1977), 'Food Supplies', in Elizabeth Stamp (ed.), *Growing Out of Poverty*, Oxford: Oxford University Press, 20–33.

Brown, Peter G. (unpublished), 'Trusteeship and Consumption', paper presented to 1994 University of Maryland Conference on 'Consumption, Global Stewardship and the Good Life'.

Brown, Peter G. (forthcoming), 'Toward an Economics of Stewardship: The Case of Climate', in *Ecological Economics*.

Bull, Hedley (1991), 'Martin Wight and the Theory of International Relations', in Martin Wight, *International Theory: The Three Traditions*, London: Leicester University Press, ix-xxiii.

Bull, Hedley (1995), *The Anarchical Society: A Study of Order in World Politics* (2nd edn), Basingstoke and London: Macmillan.

Callahan, Daniel (1971), *Ethics and Population Limitation*, New York: Population Council.

Castro, Fidel (1993), *Tomorrow is Too Late: Development and the Environmental Crisis in the Third World*, Melbourne: Ocean Press.

Chappell, Timothy (ed.) (1998), *The Philosophy of the Environment*, Edinburgh: Edinburgh University Press.

Child, Brian (1993), 'The Elephant as a Natural Resource', *Wildlife Conservation*, March/April, 60–1.

CITES (Convention on International Trade in Endangered Species of Wild Fauna and Flora, 1973) (1997), *Text of the Convention*, Geneva: CITES.

Clark, Stephen R. L. (1995), 'Environmental Ethics', in Peter Byrne and Leslie Houlden (eds), *Companion Encyclopedia of Theology*, London: Routledge, 843–68.

Cmd.12200 (1990), *The Common Inheritance*, London: HMSO.

Cmd.3789 (1997), *Eliminating World Poverty: A Challenge for the 21st Century* (White Paper on International Development), London: HMSO.

Cohen, L. Jonathan (1954), *The Principles of World Citizenship*, Oxford: Basil Blackwell.

Coleman, William (1976), 'Providence, Capitalism and Environmental Degradation: English Apologetics in an Era of Revolution', *Journal of the History of Ideas*, 37, 27–44.

Commission on Global Governance (1995), *Our Global Neighbourhood*, Oxford: Oxford University Press.

Cooper, David, E. (1992), 'The Idea of Environment', in David E. Cooper and Joy A. Palmer (eds), *The Environment in Question*, London and New York: Routledge, 165–80.

Cooper, David E., and Joy A. Palmer (eds) (1992), *The Environment in Question*, London and New York: Routledge.

Cooper, John (1980), 'Aristotle on Friendship', in A. Oksenberg Rorty (ed.), *Essays on Aristotle's Ethics*, Berkeley: University of California Press, 324–30.

Crocker, David A. (1996), 'Hunger, Capability, and Development', in William Aiken and Hugh LaFollette (eds), *World Hunger and Morality* (2nd edn), Upper Saddle River, NJ: Prentice-Hall, 211–30.

Crocker, David A. (1998), *Transitional Justice and International Civil*

Society (Working Paper #13), University Park, MD: The National Commission on Civic Renewal.

Crocker, David A., and Toby Linden (eds) (1998), *Ethics of Consumption: The Good Life, Justice and Global Stewardship*, Lanham, MD and Oxford: Rowman & Littlefield.

Crucible Group, The (1994), *People, Plants and Patents: The Impact of Intellectual Property on Biodiversity, Conservation, Trade and Rural Society*, Ottawa: International Development Research Centre.

Daly, Herman (1995), 'On Wilfred Beckerman's Critique of Sustainable Development', *Environmental Values*, 4.1, 49–55.

Delattre, Edwin (1972), 'Rights, Responsibilities and Future Persons', *Ethics*, 82, 254–8.

Descartes, René (1967), *Philosophical Works of Descartes*, trans. Elizabeth S. Haldane and G. R. T. Ross, Cambridge: Cambridge University Press.

de-Shalit, Avner (1995), *Why Posterity Matters: Environmental Policies and Future Generations*, London and New York: Routledge.

Dobson, Andrew (1996), 'Environmental Sustainabilities: An Analysis and a Typology', *Environmental Politics*, 5.3, 401–28.

Donaldson, Thomas (1992), 'Kant's Global Rationalism', in Terry Nardin and David R. Mapel (eds), *Traditions of International Ethics*, Cambridge: Cambridge University Press, 136–57.

Dower, Nigel (1994), 'The Idea of the Environment', in Robin Attfield and Andrew Belsey (eds), *Philosophy and the Natural Environment*, Cambridge: Cambridge University Press, 143–56.

Dower, Nigel (1998), *World Ethics: The New Agenda*, Edinburgh: Edinburgh University Press.

Dryzek, John S. (1987), *Rational Ecology: Environment and Political Economy*, Oxford and New York: Basil Blackwell.

Dryzek, John S. (1994), 'Green Reason: Communicative Ethics for the Biosphere', in Lori Gruen and Dale Jamieson (eds), *Reflecting on Nature: Readings in Environmental Philosophy*, New York and Oxford: Oxford University Press, 159–74.

Dryzek, John S. (1997), *The Politics of the Earth: Environmental Discourses*, Oxford: Oxford University Press.

Dubos, René (1974), 'Franciscan Conservation and Benedictine Stewardship', in David and Eileen Spring (eds), *Ecology and Religion in History*, New York: Harper & Row, 114–36.

Dubourg, Richard and David Pearce (1996), 'Paradigms for Environmental Choice: Sustainability *versus* Optimality', in Sylvie Faucheux, David Pearce and John Proops (eds), *Models of Sustainable Development*, Cheltenham: Edward Elgar, 21–36.

Durning, Alan (1991), 'Asking How Much is Enough', in Linda Starke (ed.), *State of the World 1991: A Worldwatch Institute Report on Progress Toward a Sustainable Society*, London: Earthscan, 153–69.

Dutton, Yassin (1992), 'Natural Resources in Islam', in Fazlun Khalid and

Joanne O'Brien (eds), *Islam and Ecology*, London and New York: Cassell, 51–67.

Dyson, Tim (1996), *Population and Food: Global Trends and Future Prospects*, London and New York: Routledge .

Eckersley, Robyn (1992), *Environmentalism and Political Theory: Towards an Ecocentric Approach*, London: UCL Press.

Ehrlich, Anne H., and Paul Ehrlich (1994), 'Extinction: Life in Peril', in Lori Gruen and Dale Jamieson (eds), *Reflecting on Nature: Readings in Environmental Philosophy*, New York: Oxford University Press, 335–42.

Ehrlich, Paul R. (1968), *The Population Bomb*, New York, Ballantine.

Ekins, Paul (1993), 'Making Development Sustainable', in Wolfgang Sachs (ed.), *Global Ecology*, London and Atlantic Highlands, NJ: Zed Books, 91–103.

Elliot, Robert, and Arran Gare (eds) (1983), *Environmental Philosophy*, St. Lucia, London and New York: University of Queensland Press, and Milton Keynes: Open University Press.

Epictetus (1987), extract from 'Discourses', in A.A. Long and D. Sedley (eds), *The Hellenistic Philosophers*, vol. I, Cambridge: Cambridge University Press, 364.

Esteva, Gustavo (1992), 'Development', in Wolfgang Sachs (ed.), *The Development Dictionary*, London and Atlantic Highlands, NJ: Zed Books, 6–25.

Evernden, Neil (1992), *The Social Creation of Nature*, Baltimore and London: The Johns Hopkins University Press.

Environment Digest, The (1994), No. 89/90, November/December.

Falk, Richard (1994), 'The Making of Global Citizenship', in Bart van Steenbergen (ed.), *The Condition of Citizenship*, London, Thousand Oaks, CA and New Delhi: Sage, 127–40.

Faucheux, Sylvie, David Pearce and John Proops (eds) (1996), *Models of Sustainable Development*, Cheltenham: Edward Elgar.

Ferré, Frederick (1994), 'Personalistic Organicism: Paradox or Paradigm', in Robin Attfield and Andrew Belsey (eds), *Philosophy and the Natural Environment*, Cambridge, New York and Melbourne: Cambridge University Press, 59–73.

Finkle, Jason L., and C. Alison McIntosh (eds) (1994), *The New Politics of Population: Conflict and Consensus in Family Planning*, New York and Oxford: Oxford University Press, 1994.

Foltz, Bruce V. (1984), 'On Heidegger and the Interpretation of Environmental Crisis', *Environmental Ethics*, 6, 323–38.

Forsyth, M. G., H. M. A. Keens-Soper and P. Savigear (eds) (1970), *The Theory of International Relations: Selected Texts from Gentili to Treitschke*, London: Allen and Unwin.

Fowles, John (1977), *The Magus* (revised edn), St Albans: Triad.

Fox, Matthew (1990), lecture given at St James's Church, Piccadilly, London.

Francis, David R. (1997), 'Global Crowd Control Starts to Take Effect', *Christian Science Monitor*, 89 (22 October), 1, 9.

General Synod Board for Social Responsibility (1991), *Christians and the Environment*, London: Board for Social Responsibility.

George, Susan (1976), *How the Other Half Dies*, Harmondsworth: Penguin.

George, Susan (1988), *A Fate Worse than Debt*, Harmondsworth: Penguin.

George, Susan (1992), *The Debt Boomerang: How Third World Debt Harms Us All*, London: Pluto Press.

Gilligan, Carol (1982), *In a Different Voice: Psychological Theory and Women's Development*, Cambridge, MA: Harvard University Press.

Glacken, Clarence J. (1967), *Traces on the Rhodian Shore: Nature and Culture in Western Thought from Ancient Times to the End of the Eighteenth Century*, Berkeley and London: University of California Press.

Glover, Jonathan (1995), 'The Research Programme of Development Ethics', in Martha Nussbaum and Jonathan Glover (eds), *Women, Culture and Development: A Study of Human Capabilities*, Oxford: Clarendon Press, 116–39.

Gomberg, Paul (1986), 'Self and Others in Bentham and Sidgwick', *History of Philosophy Quarterly*, 3, 437–48.

Goodin, Robert E. (1983), 'Ethical Principles for Environmental Protection', in Robert Elliot and Arran Gare (eds), *Environmental Philosophy*, St. Lucia, London and New York: University of Queensland Press, 3–20.

Goodin, Robert E. (1985), *Protecting the Vulnerable: A Reanalysis of Our Social Responsibilities*, Chicago and London: University of Chicago Press.

Goodin, Robert E. (1990), 'Property Rights and Preservationist Duties', *Inquiry*, 33, 401–32.

Goodpaster, Kenneth (1978), 'On Being Morally Considerable', *Journal of Philosophy*, 75, 308–25.

Gould, Stephen J. (1993), *Eight Little Piggies*, London: Jonathan Cape.

Granberg-Michaelson, Wesley (1992), *Redeeming the Creation: The Rio Earth Summit: Challenges for the Churches*, Geneva: WCC Publications.

Green, J.B., and M. Turner (eds) (1994), *Jesus of Nazareth: Lord and Christ: Essays on the Historical Jesus and New Testament Christology*, Grand Rapids, MI: Eerdmans.

Grubb, Michael (1989), *The Greenhouse Effect: Negotiating Targets*, London: Royal Institute of International Affairs.

Grubb, Michael (1990), *Energy Policies and the Greenhouse Effect*, Aldershot: Gower.

Grubb, Michael, and Dean Anderson (eds) (1995), *The Emerging International Regime for Climate Change: Structures and Options after Berlin*, London: Royal Institute of International Affairs.

Gruen, Lori, and Dale Jamieson (eds) (1994), *Reflecting on Nature: Readings in Environmental Philosophy*, New York: Oxford University Press.

Guha, Ramachandra (1989), 'Radical American Environmentalism and Wilderness Preservation: A Third World Critique', *Environmental Ethics*, 11, 71–83.

Guha, Ramachandra (1997), 'The Authoritarian Biologist and the Arrogance of Anti-Humanism', *The Ecologist*, 27.1, 14–19.

Guha, Ramachandra (1997), 'Radical American Environmentalism and Wilderness Preservation: a Third World Critique' (with a new Postscript) in Ramachandra Guha and Juan Martinez-Alier, *Varieties of Environmentalism: Essays North and South*, London: Earthscan, 92–108.

Gupta, Joyeeta (1995), 'The Global Environmental Facility in its North-South Context', *Environmental Politics*, 4.1, 19–43.

Habermas, Jürgen (1976), *Legitimation Crisis* (trans. Thomas McCarthy), London: Heinemann Educational.

Hale, Sir Matthew (1677), *The Primitive Origination of Mankind*, London.

Hammond, Allen L. (ed.) (1994), *World Resources, 1994–5*, Oxford and New York: Oxford University Press.

Hardin, Garrett (1971), 'The Tragedy of the Commons', in John Barr (ed.), *The Environmental Handbook: Action Guide for the UK*, London: Ballantine/Friends of the Earth, 47–65.

Hassan, M. Kamal (forthcoming), 'World-view Orientation and Ethics: A Muslim Perspective', in Azizan H. Baharuddin (ed.), *Development, Ethics and Environment*, Kuala Lumpur: Institute for Policy Research.

Hassan, Parvez (1992), 'Moving Towards a Just International Environmental Law', in Simone Bilderbeek (ed.), *Biodiversity and International Law*, Amsterdam: IOS Press.

Hayward, Tim (1995), *Ecological Thought: An Introduction*, Cambridge: Polity Press.

Hayward, Tim, and John O'Neill (eds) (1997), *Justice, Property and the Environment: Social and Legal Perspectives*, Aldershot: Ashgate.

Hayward, Tim (unpublished) 'Constitutional Environmental Rights and Liberal Democracy', paper presented at European Consortium for Political Research, University of Warwick, 1998.

Heidegger, Martin (1971), *Poetry, Language, Thought*, New York: Harper and Row.

Heidegger, Martin (1978), 'The Question Concerning Technology', in David Farrell Krell (ed.), *Martin Heidegger: Basic Writings*, London and Henley: Routledge & Kegan Paul, 283–322.

Hemingway, Ernest (1955), *For Whom the Bell Tolls*, Harmondsworth: Penguin.

Hopkins, Gerard Manley (1953), 'The Windhover', *Poems and Prose of Gerard Manley Hopkins*, selected by W. H. Gardner, Harmondsworth: Penguin.

Hösle, Vittorio (1992), 'The Third World as a Philosophical Problem', *Social Research*, 59.2, 227–62.

Houghton, J. T., L. G. Meira Filho, B. A. Callander, N. Harris, A.

Kattenberg and K. Maskell, (eds) (1996), *Climate Change 1995: The Science of Climate Change* (published for the Intergovernmental Panel on Climate Change), Cambridge: Cambridge University Press.

Howarth, Richard B. (1992), 'Intergenerational Justice and the Chain of Obligation', *Environmental Values*, 1.2, 133–40.

Hume, David (1978), *A Treatise of Human Nature*, ed. L. A. Selby-Bigge (2nd edn), Oxford: Clarendon Press.

Hurrell, Andrew and Benedict Kingsbury (eds) (1992), *The International Politics of the Environment: Actors, Interests, and Institutions*, Oxford: Clarendon Press.

Ingold, Tim (1995), 'Beyond Anthropocentrism and Ecocentrism', unpublished presentation to a Workshop on 'Ethics, Economics and Environmental Management' of the Swedish Collegium for Advanced Study in the Social Sciences, Uppsala.

International Conference on Population and Development (held Cairo, 1994) (1995), 'Program of Action', *Population and Development Review*, 21, 187–220.

International Union for the Conservation of Nature (1980), *World Conservation Strategy*, Gland (Switzerland): IUCN / UNEP / WWF.

International Union for the Conservation of Nature (1991), *Caring for the Earth: A Strategy for Sustainable Living*, Gland, Switzerland: IUCN / UNEP / WWF.

Jackson, Ben (1990), *Poverty and the Planet: A Question of Survival*, London: Penguin.

Jacobs, Michael (1991), *The Green Economy: Environment, Sustainable Development and the Politics of the Future*, London and Concord, MA: Pluto Press.

Jacobs, Michael (1995), 'Sustainable Development, Capital Substitution and Economic Humility: A Reply to Beckerman', *Environmental Values*, 4.1, 57–68.

Jacobs, Michael (1996), for the Real World Coalition, *The Politics of the Real World*, London: Earthscan.

James, David N. (1991), 'Risking Extinction: An Axiological Analysis', *Research in Philosophy and Technology*, 11, 49–63.

Jegen, Mary (1987), 'The Church's Role in Healing the Earth', in W. Granberg-Michaelson (ed.), *Tending the Garden*, Grand Rapids, MI: Eerdmans, 93–113.

Johnson, Victoria, and Robert Nurick (1995), 'Behind the Headlines: The Ethics of the Population and Environment Debate', *International Affairs*, 71.3, 547–65.

Jubilee 2000 (1998), *Fact Sheet on International Debt*, London: Jubilee 2000.

Kant, Immanuel (1970), 'Perpetual Peace: A Philosophical Sketch', in Hans Reiss (ed.), *Kant: Political Writings*, Cambridge: Cambridge University Press, 93–130.

Kates, Robert W., and Sara Millman (1990), 'On Ending Hunger: The

Lessons of History', in Lucile F. Newman (ed.), *Hunger in History: Food Shortage, Poverty and Deprivation*, Cambridge, MA: Basil Blackwell, 389–407.

Katz, Eric (1993), 'Judaism and the Ecological Crisis', in Mary Evelyn Tucker and John A. Grim (eds), *Worldviews and Ecology: Religion, Philosophy and Environment,* Lewisburg, PA: Bucknell University Press, 55–70.

Khalid, Fazlun, and Joanne O'Brien (eds) (1992), *Islam and Ecology*, London and New York: Cassell.

Kirkby, John, Phil O'Keefe and Lloyd Timberlake (eds) (1995), *The Earthscan Reader in Sustainable Development*, London: Earthscan.

Kothari, Smitu, and Pramod Parajuli (1993), 'No Nature Without Social Justice: A Plea for Cultural and Ecological Pluralism in India', in Wolfgang Sachs (ed.), *Global Ecology*, London and Atlantic Highlands, NJ: Zed Books, 224–41.

Küng, Hans, and Karl-Josef Kuschel (eds) (1993), *A Global Ethic: The Declaration of the Parliament of the World's Religions*, London: SCM Press.

Lappé, Frances Moore, and Rachel Shurman (1994), 'Taking Population Seriously', in Lori Gruen and Dale Jamieson (eds), *Reflecting on Nature: Readings in Environmental Philosophy*, New York: Oxford University Press, 328–32.

Leslie, John (1996), *The End of the World: The Science and Ethics of Human Extinction*, Routledge: London and New York.

Long, A. A., and D. Sedley (eds) (1987), *The Hellenistic Philosophers*, vol. I, Cambridge: Cambridge University Press.

Lovejoy, Arthur O. (1976), *The Great Chain of Being: A Study of the History of an Idea*, Cambridge, MA: Harvard University Press.

Lovelock, James E. (1979), *Gaia: A New Look at Life on Earth*, Oxford: Oxford University Press.

Lovelock, James E. (1988), *The Ages of Gaia: A Biography of Our Living Earth*, New York: W. W. Norton.

Luard, Evan (ed.) (1993), *Basic Texts in International Relations*, Basingstoke and London: Macmillan.

Lucas, George R., Jr (1990), 'African Famine: New Economic and Ethical Perspectives', *Journal of Philosophy*, 87, 629–41.

Luper-Foy, Steven (1992), 'Justice and Natural Resources', *Environmental Values*, 1.1, Spring, 47–64.

McIntosh, C. Alison, and Jason L. Finkle (1995), 'The Cairo Conference on Population and Development: A New Paradigm?', *Population and Development Review*, 21.2, 223–60.

McKibben, Bill (1990), *The End of Nature*, London: Viking.

McKie, Robin (1998), 'Man "Not to Blame" for Global Warming', *The Observer*, 12 April, 1.

McKie, Robin (1998), 'Solar Wind Blows Away Theories', *The Observer*, 12 April, 9.

MacLean, Douglas, and Peter G. Brown (eds) (1983), *Energy and the Future*, Totowa, NJ: Rowman & Littlefield.

Mannison, Don, Michael McRobbie and Richard Routley (eds) (1980), *Environmental Philosophy*, Monograph Series No.2, Canberra: Department of Philosophy, Australian National University.

Martinez-Alier, J. (1993), 'Distributional Obstacles to International Environmental Policy: The Failures at Rio and Prospects after Rio', *Environmental Values*, 2.2, Summer, 97–124.

Marx, Karl (1967), *Capital* (3 vols), New York: International.

Masri, Al-Hafiz B.A. (1992), 'Islam and Ecology', in Fazlun Khalid and Joanne O'Brien (eds), *Islam and Ecology*, London and New York: Cassell, 1–23.

Meadows, Donella H., Dennis L. Meadows, Jørgen Randers and William W. Behrens III (1972), *The Limits to Growth*, London and Sydney: Pan Books.

Mellanby, Kenneth (1977), 'Ecosystem', in Alan Bullock and Oliver Stalybrass (eds), *The Fontana Dictionary of Modern Thought*, London: Collins, 190.

Mellanby, Kenneth (1977), 'Environment', in Alan Bullock and Oliver Stalybrass (eds), *The Fontana Dictionary of Modern Thought*, London: Collins, 207.

Miller, Chris (1997), 'Attributing "Priority" to Habitats', *Environmental Values*, 6.3, 341–53.

Munson, Abby (1995), 'The United Nations Convention on Biological Diversity', in John Kirkby, Phil O'Keefe and Lloyd Timberlake (eds), *The Earthscan Reader in Sustainable Development*, London: Earthscan.

Murdoch, Iris (1970), *The Sovereignty of Good*, London: Routledge & Kegan Paul.

Murray, Martyn (1995), 'The Value of Biodiversity', in John Kirkby, Phil O'Keefe and Lloyd Timberlake (eds), *The Earthscan Reader in Sustainable Development*, London: Earthscan, 17–29.

Myers, Norman (ed.) (1985), *The Gaia Atlas of Planet Management*, London and Sydney: Pan Books.

Nardin, Terry, and David R. Mapel (eds) (1992), *Traditions of International Ethics*, Cambridge: Cambridge University Press.

Narveson, Jan (1983), 'On the Survival of Humankind', in Robert Elliot and Arran Gare (eds), *Environmental Philosophy*, Milton Keynes: Open University Press, 40–57.

Newman, Lucile F. (ed.) (1990), *Hunger in History: Food Shortage, Poverty and Deprivation*, Cambridge, MA: Basil Blackwell.

Newton, Lisa H., and Catherine K. Dillingham (1994), *Watersheds: Classic Cases in Environmental Ethics*, Belmont, CA: Wadsworth.

Nisbet, Robert (1980), *History of the Idea of Progress*, London: Heinemann.

Nordhaus, William (1994), *Managing the Global Commons*, Cambridge: MIT Press.

Norton, Bryan G. (ed.) (1986), *The Preservation of Species*, Princeton, NJ: Princeton University Press.

Norton, Bryan G. (1991), *Toward Unity Among Environmentalists*, New York and Oxford: Oxford University Press.

Norton, Bryan G., and Bruce Hannon (1997), 'Environmental Values: A Place-Based Approach', *Environmental Ethics*, 19.3, 227–45.

Nussbaum, Martha (1994), 'Patriotism and Cosmopolitanism', *Boston Review*, XIX.5, October/November, 2–4.

Nussbaum, Martha (1994), 'Asking the Right Questions: Martha Nussbaum Responds', *Boston Review*, XIX.5, October/November, 33–4.

Nussbaum, Martha, and Jonathan Glover (eds) (1995), *Women, Culture and Development: A Study of Human Capabilities*, Oxford: Clarendon Press.

Nylund, Are, Arne Selvik, Gunnar Skirbekk, Andreas Steigen and Audfinn Tjønneland (1992), *The Commercial Ark: A Book on Evolution, Ecology and Ethics*, Oslo: Scandinavian University Press.

O'Connor, James (1994), 'Is Sustainable Capitalism Possible?', in Martin O'Connor (ed.), *Is Capitalism Sustainable? Political Economy and the Politics of Ecology*, New York: The Guilford Press, 152–75.

O'Connor, Martin (1993), 'On the Misadventures of Capitalist Nature', *Capitalism, Nature, Socialism*, 4.3, September 7–40.

Odhiambo, Thomas R. (1997), 'Africa Beyond Famine', in Gilbert Ogutu, Pentti Malaska and Johanna Kojola (eds), *Futures Beyond Poverty: Ways and Means Out of the Current Stalemate*, Turku: World Futures Studies Federation, 157–64.

Ogutu, Gilbert, Pentti Malaska and Johanna Kojola (eds) (1997), *Futures Beyond Poverty: Ways and Means Out of the Current Stalemate*, Turku: World Futures Studies Federation.

Oksanen, Markku (unpublished), 'Privatising Genetic Resources: Biodiversity Preservation and Intellectual Property Rights', paper presented to European Consortium for Political Research Joint Sessions, University of Warwick, 1998.

O'Neill, John (1993), 'Future Generations: Present Harms', *Philosophy*, 68, 35–51.

O'Neill, John (1993), *Ecology, Policy and Politics: Human Well-Being and the Natural World*, London and New York: Routledge.

O'Neill, John (1994), 'Should Communitarians be Nationalists?', *Journal of Applied Philosophy*, 11.2, 135–43.

O'Neill, Onora (1986), *Faces of Hunger: An Essay on Poverty, Justice and Development*, London: Allen & Unwin.

Open University (1997), 'Biodiversity', in Part IV of *Environmental Ethics* (Course T861), Milton Keynes: Open University.

O'Riordan, Tim, and James Cameron (eds) (1994), *Interpreting the Precautionary Principle*, London: Cameron & May.

Page, Talbot (1983), 'Intergenerational Justice as Opportunity', in Douglas

MacLean and Peter G. Brown (eds), *Energy and the Future*, Totowa, NJ: Rowman & Littlefield, 1983, 38–58.

Palmer, Clare (1992), 'Stewardship: A Case Study in Environmental Ethics', in J. Ball, M. Goodhall, C. Palmer and J. Reader (eds), *The Earth Beneath*, SPCK: London, 67–86.

Parfit, Derek (1983), 'Energy Policy and the Further Future: The Social Discount Rate', in Douglas MacLean and Peter G. Brown (eds), *Energy and the Future*, Totowa, NJ: Rowman & Littlefield, 166–79.

Parfit, Derek (1984), *Reasons and Persons*, Oxford: Clarendon Press.

Partridge, Ernest (1981), 'Why Care About the Future?', in Ernest Partridge (ed.), *Responsibilities to Future Generations: Environmental Ethics*, Buffalo: Prometheus, 203–20.

Partridge, Ernest (ed.) (1981), *Responsibilities to Future Generations: Environmental Ethics*, Buffalo: Prometheus.

Passmore, John (1974), *Man's Responsibility for Nature*, London: Duckworth.

Passmore, John (1995), 'The Preservationist Syndrome', *Journal of Political Philosophy*, 3.1, 1–22.

Paul VI, Pope (1967), Encyclical Letter *Populorum Progressio* (on Fostering the Development of Peoples), London: Catholic Truth Society.

Pearce, David, Anil Markandya, Edward B. Barbier (1989), *Blueprint for a Green Economy*, London: Earthscan.

Pearce, David, Edward Barbier, Anil Markandya, Scott Barrett, R. Kerry Turner and Timothy Swanson (1991), *Blueprint 2: Greening the World Economy*, London: Earthscan.

Percy, Steve (1996), 'Arid Aral', *The New Internationalist*, 277, March, 4.

Pickering, Kevin T., and Lewis A. Owen (1997), *An Introduction to Global Environmental Issues* (2nd edn), London and New York: Routledge.

Piel, Gerard (ed.) (1990), *The World of René Dubos*, New York: Henry Holt.

Pirages, Dennis Clark (ed.) (1977), *The Sustainable Society: Implications for Limited Growth*, New York and London: Praeger.

Pogge, Thomas (1992), 'Cosmopolitanism and Sovereignty', *Ethics*, 103, 48–75.

Popper, Karl (1963), *Conjectures and Refutations*, London: Routledge & Kegan Paul.

Porter, Roy (1996), 'The End is Nigh', *The Observer*, 14 April (Review Section), 14.

Postel, Sandra (1996), 'Forging a Sustainable Water Strategy', in Lester R. Brown et al. (eds), *State of the World, 1996: A Worldwatch Institute Report on Progress Towards a Sustainable Society*, London: Earthscan, 40–59.

Prest, Michael (1997), 'Water, Water, Nowhere?', *Review*, London: Rio Tinto, 15–19.

Pretty, Jules (1998), 'Feeding the World?' *Splice*, 4.6, August/September, 4–6.

Program of Action of the International Conference on Population and Development, Cairo, 1994 (1995), *Population and Development Review*, 21, Part 1, 187–213, and Part 2, 437–61.

Rawls, John (1972), *A Theory of Justice*, London, Oxford and New York: Oxford University Press.

Rawls, John (1993), *Political Liberalism*, New York and Chichester: Columbia University Press.

Redclift, Michael (1984), *Development and the Environmental Crisis: Red or Green Alternatives?*, London and New York: Methuen.

Redclift, Michael, and Ted Benton (eds) (1994), *Social Theory and the Global Environment*, London and New York: Routledge.

Reiss, Hans (ed.) (1970), *Kant: Political Writings*, Cambridge: Cambridge University Press.

Roche, Pete (1997–8), 'The Atlantic Frontier Debate: Time for Ecological Limits?', *New Ground*, Winter, 14–16.

Rolston, Holmes, III (1986), 'Duties to Endangered Species', in Holmes Rolston III, *Philosophy Gone Wild: Essays in Environmental Ethics*, Buffalo: Prometheus, 206–20.

Rolston, Holmes, III (1996), 'Feeding People versus Saving Nature', in William Aiken and Hugh LaFollette (eds), *World Hunger and Morality* (2nd edn), Upper Saddle River, NJ: Prentice-Hall, 244–63.

Rolston, Holmes, III (1998), 'Nature for Real: Is Nature a Social Construct?', in Timothy Chappell (ed.), *The Philosophy of the Environment*, Edinburgh: Edinburgh University Press, 38–64.

Routley, Richard, and Val Routley (1980), 'Human Chauvinism and Environmental Ethics', in Don Mannison, Michael McRobbie and Richard Routley (eds), *Environmental Philosophy*, Monograph Series No.2, Canberra: Department of Philosophy, Australian National University, 96–189.

Ryberg, Jesper (1997), 'Population and Third World Assistance: A Comment on Hardin's Lifeboat Ethics', *Journal of Applied Philosophy*, 14.3, 207–19.

Sachs, Wolfgang (1993), 'Global Ecology and the Shadow of "Development" ', in Wolfgang Sachs (ed.), *Global Ecology*, London and Atlantic Highlands, NJ: Zed Books, 3–21.

Sachs, Wolfgang (ed.) (1993), *Global Ecology*, London and Atlantic Highlands, NJ: Zed Books.

Sagoff, Mark (1974), 'On Preserving the Natural Environment', *Yale Law Journal*, 84, 205–67.

Sagoff, Mark (1998), 'Carrying Capacity and Ecological Economists', in David A. Crocker and Toby Linden (eds), *Ethics of Consumption: The Good Life, Justice, and Global Stewardship*, Lanham, MD and Oxford: Rowman & Littlefield, 28–52.

Sale, Kirkpatrick (1985), *Dwellers in the Land: The Bioregional Vision*, San Francisco: Sierra Club Books.

Santmire, H. Paul (1985), *The Travail of Nature: The Ambiguous Ecological Promise of Christian Theology*, Philadelphia, PA: Fortress Press.

Sarre, Philip, and John Blunden (1995), *An Overcrowded World? Population, Resources and the Environment*, Oxford: Oxford University Press.

Schell, Jonathan (1982), *The Fate of the Earth*, New York: Alfred A. Knopf.

Schmidtz, David (1997), 'Why Preservationism Doesn't Preserve', *Environmental Values*, 6.3, 327–39.

Sen, Amartya (1981), *Poverty and Famines: An Essay on Entitlement and Deprivation*, New York: Oxford University Press.

Shiva, Vandana (1991), *The Violence of the Green Revolution: Third World Agriculture, Ecology and Politics*, London and Atlantic Highlands, NJ: Zed Books.

Shiva, Vandana (1993), *Monocultures of the Mind: Perspectives on Biodiversity and Biotechnology*, London: Zed Books, and Penang: Third World Network.

Shiva, Vandana (1997), *Biopiracy: The Plunder of Nature and Knowledge*, Boston, MA: South End Press.

Shrader-Frechette, K. S., and E. D. McCoy (1994), 'Biodiversity, Biological Uncertainty, and Setting Conservation Priorities', *Biology and Philosophy*, 9.2, 167–95.

Shue, Henry (1992), 'The Unavoidability of Justice', in Andrew Hurrell and Benedict Kingsbury (eds), *The International Politics of the Environment: Actors, Interests, and Institutions*, Oxford: Clarendon Press, 373–97.

Shue, Henry (1995), 'Ethics, the Environment and the Changing International Order', *International Affairs*, 71.3, 453–61.

Shue, Henry (unpublished), 'Equity in an International Agreement on Climate Change', paper presented to IPCC workshop on 'Equity and Social Considerations Related to Climate Change', Nairobi, 1994.

Sikora, R. I., and Brian Barry (eds) (1978), *Obligations to Future Generations*, Philadelphia: Temple University Press.

Simon, Julian (1981), *The Ultimate Resource*, Oxford: Martin Robertson.

Singer, Peter (1976), *Animal Liberation: A New Ethics for Our Treatment of Animals*, London: Jonathan Cape.

Sober, Elliot (1986), 'Philosophical Problems for Environmentalism', in B. G. Norton (ed.), *The Preservation of Species*, Princeton, NJ: Princeton University Press, 173–94.

Spoor, Max (unpublished), 'Political Economy of the Aral Sea Crisis', paper presented at international Conference on 'Transformation Processes in Eastern Europe', Amsterdam, March 1997.

Spring, David, and Eileen Spring (eds) (1974), *Ecology and Religion in History*, New York: Harper & Row.

Starke, Linda (ed.) (1991), *State of the World 1991: A Worldwatch Institute Report on Progress Toward a Sustainable Society*, London: Earthscan.

Stone, Christopher D. (1993), *The Gnat is Older than Man: Global Environment and Human Agenda*, Princeton, NJ: Princeton University Press.

Talbot, Carl (1997), 'Environmental Justice', *Encyclopedia of Applied Ethics*, vol. 2, San Diego: Academic Press, 93–103.

Taylor, Paul (1986), *Respect for Nature: A Theory of Environmental Ethics*, Princeton, NJ: Princeton University Press.

Thomas, Keith (1983), *Man and the Natural World: A History of the Modern Sensibility*, New York: Pantheon Books.

Thomas, Neil (1996), 'Who Defused the Population Bomb?', *Planet: The Welsh Internationalist*, 116, 85–92.

Thompson, Janna (1992), *Justice and World Order: A Philosophical Inquiry*, London and New York: Routledge.

Thompson, Janna (unpublished), 'Toward a Green World Order: Environment and World Politics'.

Thompson, Thomas H. (1981), 'Are We Obligated to Future Others?', in Ernest Partridge (ed.), *Responsibilities to Future Generations: Environmental Ethics*, Buffalo: Prometheus, 195–202.

Thoreau, Henry David (1968), *Walden*, London: Dent.

Tiffen, Mary, Michael Mortimore and Francis Gichuki (1994), *More People, Less Erosion: Environmental Recovery in Kenya*, Chichester: Wiley.

Toulmin, Stephen (1990), *Cosmopolis, The Hidden Agenda of Modernity*, New York: Free Press.

Tucker, Aviezer (1994), 'In Search of Home', *Journal of Applied Philosophy*, 11.2, 181–7.

Tucker, Mary Evelyn, and John A. Grim (eds) (1993), *Worldviews and Ecology: Religion, Philosophy and Environment*, Lewisburg, PA: Bucknell University Press.

Tudge, Colin (1995), *The Day Before Yesterday*, London: Jonathan Cape.

Turner, B. L., II, Roger E. Kasperson and William B. Meyer (1990), 'Two Types of Global Environmental Change', *Global Environmental Change*, 1.1, 15–17.

United Nations (1986), *Declaration on the Right to Development*, New York: United Nations.

United Nations Conference on Environment and Development (1992), 'Rio Declaration on Environment and Development', in Wesley Granberg-Michaelson, *Redeeming the Creation: The Rio Earth Summit: Challenges for the Churches*, Geneva: WCC Publications, 86–90.

United Nations Environment Programme/World Wide Fund for Nature/World Conservation Union (1991), *Caring for the Earth*, London: Earthscan.

United Nations Environmental Programme (1992), 'The State of the Global Environment', *Our Planet*, 4.2, 4–9.

Vallely, Paul, and Ian Linden (1997), 'Doing Battle with Globalisation', 1, 'Are We Masters or Servants?', *The Tablet*, 9 August, 1004–5.

Vandeloo, Ted (unpublished), 'Water, Ethics and the Global Village', paper presented at 1996 Aberdeen conference on 'Ethics, Development and Global Values'.

VanDeVeer, Donald (1979), 'Interspecific Justice', *Inquiry*, 22, 55–79.

Vidal, John (1998), 'Baptism of Fire', *The Guardian* (Society Section), 20 May, 4–5.

Vitoria, Francisco de (1993), 'On the Indians', in Evan Luard (ed.), *Basic Texts in International Relations*, Basingstoke and London: Macmillan, 145–9.

Waddell, Helen (1934), *Beasts and Saints*, London: Constable.

Wallace-Hadrill, D. S. (1968), *The Greek Patristic View of Nature*, Manchester: Manchester University Press, and New York: Barnes and Noble.

Walzer, Michael (1995), 'Introduction', in Michael Walzer (ed.), *Toward a Global Civil Society*, Providence, RI and Oxford: Berghahn, 1–4.

Walzer, Michael (ed.) (1995), *Toward a Global Civil Society*, Providence, RI and Oxford: Berghahn.

Walzer, Michael (1995), 'The Civil Society Argument', in Ronald Beiner (ed.), *Theorizing Citizenship*, Albany, NY: State University of New York Press, 153–74.

Watson, Robert T., Marafu C. Zinyowera and Richard H. Moss (eds) (1996), *Climate Change 1995: Impacts, Adaptations and Mitigation of Climate Change: Scientific-Technical Analyses*, Cambridge: Cambridge University Press.

Weiss, Edith Brown (1989), *In Fairness to Future Generations: International Property Law, Common Patrimony and Intergenerational Equity*, New York: United Nations University, Japan and Transnational Publishers.

Wenz, Peter S. (1988), *Environmental Justice*, Albany, NY: State University of New York Press.

Wersal, Lisa (1995), 'Islam and Environmental Ethics: Tradition Responds to Contemporary Challenges', *Zygon*, 30, 451–9.

White, Tyrene (1994), 'Two Kinds of Production: The Evolution of China's Family Planning Policy in the 1980s', in Jason L. Finkle and C. Alison McIntosh (eds), *The New Politics of Population: Conflict and Consensus in Family Planning*, New York and Oxford: Oxford University Press, 137–58.

Wight, Martin (1991), *International Theory: The Three Traditions*, London: Leicester University Press.

Wilkins, Barry (1996), 'Civil Society', in Michael Payne (ed.), *A Dictionary of Cultural and Critical Theory*, Oxford: Blackwell, 102–3.

Williams, Mary B. (1978), 'Discounting versus Maximum Sustainable Yield', in R. I. Sikora and Brian Barry (eds), *Obligations to Future Generations*, Philadelphia: Temple University Press.

Wiwa, Ken (1998), 'Prime Mover', *The Guardian* (Society Section), 20 May, 4–5.

Wood, Paul M. (1997), 'Biodiversity as a Source of Biological Resources: A New Look at Biodiversity Values', *Environmental Values*, 6.3, 251–68.

World Bank, The (1992), *World Development Report, 1992*, New York: Oxford University Press.

World Commission on Environment and Development (1987), *Our Common Future*, Oxford: Oxford University Press.

World Health Organisation (1994), *Operation and Maintenance of Urban Water Supply and Sanitation Systems: A Guide for Managers*, Geneva: WHO.

Worster, Donald (1993), 'The Shaky Ground of Sustainability', in Wolfgang Sachs (ed.), *Global Ecology: A New Arena of Political Conflict*, London and Atlantic Highlands, NJ: Zed Books, 132–45.

Yamin, Farhana (1995), 'Biodiversity, Ethics and International Law', *International Affairs*, 71.3, 529–46.

Yearley, Steven (1996), *Sociology, Environmentalism, Globalization: Reinventing the Globe*, London, Thousand Oaks, CA, and New Delhi: Sage Publications.

Young, Michael D. (1993), *For Our Children's Children: Some Practical Implications of Inter-Generational Equity and the Precautionary Principle*, Canberra: Resource Assessment Commission.

INDEX

Learning Resources Centre